The Story of the (

Frederick Wilkinson

Alpha Editions

This edition published in 2024

ISBN : 9789362991461

Design and Setting By
Alpha Editions
www.alphaedis.com
Email - info@alphaedis.com

As per information held with us this book is in Public Domain.
This book is a reproduction of an important historical work. Alpha Editions uses the best technology to reproduce historical work in the same manner it was first published to preserve its original nature. Any marks or number seen are left intentionally to preserve its true form.

Contents

PREFACE. ....- 1 -

CHAPTER I. ....- 2 -

ORIGIN, GROWTH, AND CHIEF CULTIVATED SPECIES OF COTTON PLANT. ....- 2 -

CHAPTER II. ....- 18 -

COTTON-PLANT DISEASES AND PESTS. ....- 18 -

CHAPTER III. ....- 22 -

CULTIVATION OF THE COTTON PLANT IN DIFFERENT COUNTRIES. ....- 22 -

CHAPTER IV. ....- 39 -

THE MICROSCOPE AND COTTON FIBRE. ....- 39 -

CHAPTER V. ....- 44 -

PLANTATION LIFE AND THE EARLY CLEANING PROCESSES. ....- 44 -

CHAPTER VI. ....- 52 -

MANIPULATION OF COTTON IN OPENING, SCUTCHING, CARDING,

DRAWING, AND FLY-FRAME MACHINES. ....- 52 -

CHAPTER VII. ....- 71 -

EARLY ATTEMPTS AT SPINNING, AND EARLY INVENTORS. ....- 71 -

CHAPTER VIII. ....- 80 -

FURTHER DEVELOPMENTS— ARKWRIGHT AND CROMPTON. ....- 80 -

CHAPTER IX. ....- 93 -

THE MODERN SPINNING MULE. ....- 93 -

CHAPTER X. ....- 101 -

OTHER PROCESSES IN COTTON SPINNING. ....- 101 -

CHAPTER XI. ....- 111 -

DESTINATION OF THE SPUN YARN. ....- 111 -

PREFACE.

In collecting the facts which will be found in this Story of the Cotton plant, the author has of necessity had to consult many books. He is especially indebted to Baines' "History of the Cotton Manufacture," French's "Life and Times of Samuel Crompton," Lee's "Vegetable Lamb of Tartary," Report of the U. S. A. Agricultural Department on "The Cotton Plant," and The American Cotton Company's Booklet on the Cylindrical Bale.

Mr. Thornley, spinning master at the Technical School, Bolton, has from time to time offered very important suggestions during the progress of this little work. The author is also deeply indebted to the late Mr. Woods of the Technical School, Bolton, who was good enough to photograph most of the pictures which illustrate this book, and without which it would have been impossible to make the story clear.

For permission to reproduce Fig. 3, the thanks of the author are due to Messrs. Sampson Low and Co., for Fig. 4, to Messrs. Longmans, Green and Co. For Figs. 5, 8, 9, 13, and 36, to Messrs. Dobson and Barlow, Ltd., Bolton. For Fig. 7, viz., the Longitudinal and Transverse Microphotographs of Cotton Fibre, the author is much indebted to Mr. Christie of Mark Lane, London, who generously photographed them especially for this work. For Fig. 23, I am obliged to Mr. A. Perry, Bolton.

FRED. WILKINSON.

TEXTILE AND ENGINEERING SCHOOL,
BOLTON.

CHAPTER I.

ORIGIN, GROWTH, AND CHIEF CULTIVATED SPECIES OF COTTON PLANT.

In the frontispiece of this little work is a picture of a cotton field showing the plants bearing mature pods which contain ripe fibre and seed, and in Fig. 2 stands a number of bobbins or reels of cotton thread, in which there is one having no less than seventeen hundred and sixty yards of sewing cotton, or one English mile of thread, on it. As both pictures are compared there appears to be very little in common between them, the white fluffy feathery masses contained in the pods shown in the one picture, standing in strange contrast to the strong, beautifully regular and even threads wound on the bobbins pictured in the other.

From cotton tree to cotton thread is undoubtedly a far cry, but it will be seen further on that the connection between the two is a very real and vital one.

Now it is the main purpose of this book to unfold the wonderful story of the plant, and to fill in the details of the gap from tree to thread, and to trace the many changes through which the beautiful downy cotton wool passes before it arrives in the prim looking state of thread ready alike for the sewing machine or the needle of a seamstress.

FIG. 2.—Bobbins of cotton thread.

Remembering that the great majority of the readers of this little book must of necessity be quite unaccustomed to trade terms and technical expressions, the author has endeavoured to present to his readers in untechnical language a simple yet truthful account of the many operations and conditions through which cotton is made to pass before reaching the final stages.

Nature provides no lovelier sight than the newly opened capsules containing the pure white and creamy flocculent masses of the cotton fibre as they hang from almost every branch of the tree at the end of a favourable season.

And how strange is the story of this plant as we look back through the centuries and listen to the myths and fables, almost legion, which early historians have handed down to us or imaginative travellers have conceived. There is, however, every reason to believe that in the far distant ages of antiquity this plant was cultivated, and yielded then, as it does now, a fibre from which the inhabitants of those far-off times produced material with which to clothe their bodies.

It will not be considered out of place if some of the early beliefs which obtained among the peoples of Western Asia and Europe for many years are related.

Like many other things the origin of the Cotton plant is shrouded in mystery, and many writers are agreed that it originally came from the East, but it will be seen later on that equally strong claims can be presented from other countries in the Western Hemisphere. Many of us have been amused at the curious ideas which people, say of a hundred years ago, had of the Coral Polyp.

Even to-day children may be heard singing in school,

"Far adown the silent ocean
Dwells the coral *insect* small"!

Not a few of the early naturalists believed that the Coral was a plant and while living in the sea water it was soft, and when dead it became hard!

We smile at this, of course, but it was not until actual investigation on the spot, as to what the Coral was, that the truth came out.

It was then discovered to be an animal and not a plant, and that during life its hard limy skeleton was covered by soft muscular tissue, which, when decomposing, was readily washed away by the sea, leaving the hard interior exposed as coral.

When the absurd beliefs are read which found credence among all classes of the people during the middle ages, and down even to the end of the seventeenth century, as to what the cotton boll or pod was, the reader is inclined to rub his eyes and think surely he must be reading "Baron Munchausen" over again, for a nearer approach to the wonderful statements of that former-fabled traveller it would be difficult to find than the simple crude conceptions which prevailed of the growth, habits, and physical characteristics of the Cotton plant.

The subject of the early myths and fables of the plant in question has been very fully treated by the late Mr. Henry Lee, F. L. S., who was for a time at the Brighton Aquarium. His book, the "Vegetable Lamb of Tartary," shows indefatigable research for a correct explanation of the myth, and after a strictly impartial inquiry he comes to the conclusion that all the various phases which these fabulous concoctions assumed, had their beginnings in nothing more or less than the simple mature pod of the Cotton plant.

It will not be necessary to consider here more than one or two of these very curious beliefs about cotton. By some it was supposed that in a country which went by the name "The Tartars of the East," there grew a wonderful tree which yielded buds still more wonderful. These, when ripe, were said to burst and expose to view tiny lambs whose fleeces gave a pure white wool which the natives made into different garments.

By and by, a delightfully curious change took place, and it is found that the fruit which was formerly said to have the little lamb within, was now changed into a live lamb attached to the top of the plant. Mr. Lee says: "The stem or stalk on which the lamb was suspended above the ground, was sufficiently flexible to allow the animal to bend downward, and browse on the herbage within its reach. When all the grass within the length of its tether had been consumed, the stem withered and the plant died. This plant lamb was reported to have bones, blood, and delicate flesh, and to be a favourite food of wolves, though no other carnivorous animal would attack it."

Planta Tartarica Boromez

FIG. 3.—The vegetable lamb of Tartary.

In Fig. 3 is shown Joannes Zahn's idea of what this wonderful "Barometz or Tartarian lamb" was like. Now, mainly through an imaginative Englishman named Sir John Mandeville, who lived in the reign of Edward III., did this latter form of the story find its way into England.

This illustrious traveller left his native country in 1322, and for over thirty years traversed the principal countries of Europe and Asia. When he came home he commenced to write a history of his remarkable travels. In these are found references to the Cotton plant, and so curious an account does he give of it, that it has been considered worth reproduction in his own words: "And there growethe a maner of Fruyt, as though it weren Gourdes: and whan ther been rype men kutten hem ato, and men fynden with inne a lyttle Best, in Flesche, in Bon and Blode, as though it were a lytylle Lomb with outen Wolle. And men eten both the Frut and the Best; and that is a great Marveylle. Of that Frute I have eaten; alle thoughe it were wondirfulle, but that I knowe well that God is Marveyllous in his Werkes."

No wonder that many accepted his account of the "Vegetable Lamb" without question. When a nobleman of the reputation of Sir J. Mandeville stated that he had actually eaten of the fruit of the Cotton, was there any need for further doubt?

It appears, however, that contemporary with Mandeville was another traveller, an Italian Friar, named Odoricus, who also had travelled in Asia and heard of the plant which yielded cotton. He, too, fell a prey to the lamb theory. Many other writers and travellers followed, all more or less believing in the plant animal theory. However, in 1641, Kircher of Avignon in describing cotton declared it to be a plant. And so the story for years passed through many changes. First one would assert what he considered to be the right solution, and this was immediately challenged by the next investigator, so that assertion and contradiction followed each other in quick succession.

In 1725, however, a German doctor named Breyn communicated with the Royal Society on the subject of the "Vegetable lamb," emphatically stating the story to be nothing more or less than a fable. He very naïvely remarked that "the work and productions of nature should be discovered, not invented," and he threw doubts as to whether those who had written about the mythical lamb had ever seen one.

When the writings and dissertations of Mandeville, Odoricus and others are carefully considered, these conclusions force themselves upon us: that direct personal observation must have played a very minor part in the attempt to get at the truth in connection with the origin and growth of the Cotton plant.

Their statements stand in very sharp contrast with those of writers who lived before the Christian era commenced. Of these, mention must be made of Herodotus, surnamed the *Father of History*.

This celebrated Greek historian and philosopher was born, B.C. 484, in Halicarnassus in Greece. In his book of travels he speaks of the Cotton plant. It appears, mainly owing to the tyrannical government of Lygdamis, he left his native land and travelled in many countries in Europe, Asia, and Africa. He appears to have at least determined, that he would only write of those things of which he had intimate knowledge, and would under no circumstances take for granted what he could not by personal observation verify for himself. In speaking of India and the Cotton plant, he says: "The wild trees in that country bear for their fruit fleeces surpassing those of sheep in beauty and excellence, and the natives clothe themselves in cloths made therefrom." In another place he refers to a present which was sent by one of the kings of Egypt, which was padded with cotton. He also describes a machine for separating the seed from the fibre or lint. Compared with our modern gins, as they are called, this machine was exceedingly primitive and simple in construction.

There is not the slightest doubt that the first reliable information of the physical characters of the fibre and its uses was conveyed into Europe by the officers of the Emperor Alexander. One of his greatest Admirals, named Nearchus, observed the growth of cotton in India, and the use to which it was put, especially the making of sheets, shirts and turbans.

Perhaps one of the most careful observers that lived before the Christian era commenced, was Theophrastus, who wrote some strikingly correct things about the Cotton plant of India three centuries before Christ!

In describing the tree he said it was useful in producing cotton which the Indians wove into garments, that it was not unlike the dog rose, and that the leaves were somewhat like the leaves of the mulberry tree. The cultivation of the plant was also very correctly noted as to the rows in which the cotton seeds were placed, and as to the distances to which these rows were set. According to Dr. Royle, however, reference is made to cotton in the "Sacred Institutes of Manu" so frequently that the conclusion is admitted that cotton must have been in frequent use in India at that time, which was 800 B.C.

As was to be expected, Persia very early had cottons and calicoes imported from India. In the sixth verse of the first chapter of Esther definite reference is made to the use to which cotton was put at the feasts which King Ahasuerus gave about 519 B.C. "White, green, and blue hangings" are said to have been used on this occasion, and from authorities who have specially investigated this subject, we are told that the hangings mentioned were simply white and blue striped cottons. This would also confirm the statement that dyeing is one of the oldest industries we have.

It appears that the Greeks and Romans in good time learned to value goods made of cotton, and soon followed the Oriental custom of erecting awnings or coverings for protection from the sun's rays. The Emperor Cæsar is said to have constructed a huge screen extending from his own residence along the Sacred Way to the top of the Capitoline Hill. The whole of the Roman Forum was also covered in by him in a similar way. Coverings for tents, sail cloth made from cotton, and fancy coverlets were also in use among the people of these stirring times.

And now comes the important question: Was cotton indigenous to India in these very early times? and was it carried and afterwards planted in Egypt, Africa, and America?

As an attempt is made to successfully answer this question, our minds are thrown back to the time when Christopher Columbus, a Genoese, having heard of India, desired to find a new way to that country. Comparatively poor himself, he was unable to equip an expedition, and laid his scheme before the Council of Genoa. They declined to have anything to do with it, and he is found next presenting his case to the King of Portugal. Here he alike failed, and he ultimately applied to the King and Queen of Spain, when he met with success.

The 3rd of August, 1492, found him fully equipped with two ships, and on his way west to find a new way to India. He first touched the Bahamas thirty days after setting sail from Europe, and to his astonishment he was met by the natives, who came out to meet him in canoes, bringing with them cotton yarn and thread for the purpose of barter. In Cuba he was surprised to find hammocks made from cotton cord in very general use. What Columbus observed in the West Indies as to the growth and manufacture of cotton, was found afterwards to be by no means confined to these islands, but that in South and Central America the natives were quite accustomed both to the growth and manufacture of cotton.

Indisputable evidence can be presented to prove that the ancient civilisations of Mexico, Peru, and Central America, were well acquainted with cotton. When Peru was subjugated in 1522 by Pizarro, the manufacture of cotton was in a flourishing condition.

Similarly when Mexico fell into the hands of Cortez in 1519, he too found that the use of cotton was very general. So delighted was he at what he saw of the quality and beauty of their manufactured goods, that he had no hesitation in dispatching to Europe a present consisting of mantles, to the Emperor Charles V.

Five years after Columbus started on his momentous voyage, another expedition under Vasco da Gama set out from the Tagus to make the voyage to India by the way of the Cape of Good Hope.

Immediately Gama had safely reached India, there were others who quickly desired to follow, and in 1516 another adventurous Spaniard on his way to India called at S. Africa, and found the natives wearing garments made of cotton.

There is therefore no reason to question the statement which has repeatedly been made, that at least three centres are known in which the Cotton plant from very early times has been indigenous, and that the peoples of these countries were well acquainted with the property and uses of the cotton wool obtained from the plant. An average of more than 1,000,000 bales, each weighing 500 lbs., are exported from Egypt every year, and the question has been raised whether the cultivation of the plant in Egypt can be said to date far back. This is not so. The fibre almost exclusively used by the ancient Egyptians was flax, and the nature of the garments covering the mummies of the ancient Egyptians has been satisfactorily decided by the microscope. It is very probable that the cultivation of the plant at the beginning of the thirteenth century was carried on purely for the purpose of ornamental gardening, and even when the seventeenth century was fairly well advanced, the Egyptians still imported cotton.

The nineteenth century, however, has seen important developments in the cultivation of cotton in Egypt, and now the position attained by this country is only outdistanced by the United States and India.

The Botany of Cotton.—Botanists tell us that the vegetable kingdom is primarily divided into three great classes—viz., (1) Dicotyledons; (2) Monocotyledons; and (3) Acotyledons.

Now these names solely refer to the nature and form of the seeds produced by the plants, and by the first it is understood that a single seed is divisible into two seed lobes in developing. In the case of the second, the seed is formed only of one lobe, and in the third the seed is wanting as a cotyledon, but the method of propagation is carried on by what are called spores. We have examples of the last-named class in the ferns, lycopods and horsetail plants. The first two of the above-named classes have been well called Seed plants. These are again broken up into divisions, to which the name Natural Orders has been given. Most of us know, as the following are examined, Anemone, Buttercup, Marsh Marigold, Globe Flower, and Larkspur, that they have the same general structural arrangement, but in many particulars they differ. Thus these natural orders are again subdivided into genera, and a still further subdivision into species is made.

The Cotton plant is put in the genus *Gossypium*, which is one falling into the natural order *Malvacæ*, and which is one of a very large number forming an important division of the dicotyledons where the stamens are found to be inserted below the pistil, and where the corolla is composed of free separate petals, and where the plant has a flower bearing both calyx and corolla. So far as numbers are concerned, the Malvacæ cannot be said to be important, but few genera being known to fall into this order. Three are familiar at least—viz., the Marsh Mallow, which was formerly used a great deal in making ointment; the Musk Mallow, and the Tree Mallow. The most important genus in this order is the Gossypium. This name was given to the Cotton plant by Pliny, though the reasons for so doing are not clear. Very many species are known to exist at the present time, and this is not to be wondered at, when the area in which the plant is cultivated is so vast, and coupled with the fact that the plant is susceptible to the slightest change and "sports" most readily.

Differences of soil, climate, position with regard to the sea board, and variations in the method of cultivation could only be expected to result in the species being exceedingly numerous. It is not surprising, therefore, to find that no two botanists agree as to the number of species comprising the Gossypium family. A list, however, of the commoner varieties found in various cotton-growing areas of the globe will be given, but before doing so, it is deemed advisable to give a general botanical description of the plant.

The Gossypium is either herbaceous, shrubby, or treelike, varying in height from three to twenty feet. In some cases it is perennial; in most, as in the cultivated species, it is an annual or biennial. A few examples are noted for the vast number of hairs found everywhere on the plant, and on almost every part of the plant also, there may be observed black spots or glands. Usually the stem is erect, and as a rule the Cotton plant in form is not unlike the fir tree, that is, its lower branches are wider than those above, and this gradual tapering extends to the top of the tree. In consequence of this it is said to be *terete*. The leaves are alternate, veined and petiolated, that is, they have a leaf stalk connecting leaf and stem. In shape the leaves are cordate or heart-shaped, as well as sub-cordate, and the number of lobes found in the leaf varies from three to seven. The stipules or little appendages found on the petioles, resembling small leaves in appearance and texture, are generally found in pairs. The calyx is cup-shaped, and the petals of the flower are very conspicuous, and vary in colour according to the species, being brownish-red, purple, rose-coloured, and yellow. The petals, five in number, are often joined together at the base. The ovary is sessile, that is, it directly rests upon the main stem, and is usually three to five celled. The pod or capsule, which contains the seeds and cotton fibre, when ripe splits into valves, which vary in number from three to five. A characteristic feature of the pod is the sharp top point formed by the meeting of the pointed valves. The seeds are numerous and very seldom smooth, being usually thickly covered with fibrous matter known as raw cotton. As is well known, the wind performs a very important function in the dispersal of seeds. It is clear that when a seed is ready to be set free, and is provided by a tuft of hair, such as is seen on the cotton seed, dandelion and willow herb, it becomes a very easy matter for it to be carried ever so far, when a good breeze is blowing. Most of us have blown, when children, at the crown of white feathery matter in the dandelion, and have been delighted to see the tiny parachutes carrying off its tiny seed to be afterward deposited, and ultimately take root and appear as a new plant. Much in the same way, before it was cultivated, the Cotton plant perpetuated its own species. It should be added that the root of the Cotton plant is tap shaped, and penetrates deeply into the earth.

It would be well nigh impossible to enumerate all the species which are now known in the Cotton plant family, and it is not proposed here to describe more than the principal types of the Gossypium. In a report prepared by Mr. Tracy of Mississippi, U. S. A., no less than one hundred and thirty varieties of American cotton are given. He says: "The word 'variety' refers exclusively to the various forms and kinds which are called varieties by cotton planters, and is not restricted to the more marked and permanent types which are recognised by botanists. Of botanical varieties there are but few, while of agricultural varieties there are an almost infinite number, and the names under which the agricultural varieties are known are many times greater than the recognisable forms." The Cotton plant most readily responds to any changes of climate, methods of cultivation, change of soil or of fertilizers. So that it is easy to understand in a plant so susceptible and prone to vary as is the cotton, that new species may in a few years be brought into existence, and especially by means of proper selection of the seed, and careful cultivation.

The chief commercial types of *Gossypium* are—1. *Barbadense*; 2. *Herbaceum*; 3. *Hirsutum*; 4. *Arboreum*; 5. *Neglectum*; 6. *Peruvianum*.

Gossypium Barbadense.—The fine long silky fibres of commerce are all derived from this species. It is indigenous to a group of the West Indian Islands named the Lesser Antilles. It gets its name from Barbadoes, one of the West Indies. At the present time it is cultivated throughout the Southern States of North America which border on the sea, in most of the West Indian Islands, Central America, Western Africa lying between the tropics, Bourbon, Egypt, Australia, and the East Indies. There is no doubt that the plant comes to its highest and most perfect state of cultivation when it is planted near the sea. Dr. Evans says: "It may be cultivated in any region adapted to the olive and near the sea, the principal requisite being a hot and humid atmosphere, but the results of acclimatisation indicate that the humid atmosphere is not entirely necessary if irrigation be employed, as this species is undoubtedly grown extensively in Egypt." The height of this species varies from 3 to 4 feet if cultivated as an annual, and from 6 to 8 feet if allowed to grow as a perennial. When in full leaf and flower, it is a most graceful-looking plant. Yarns having the finest counts, as they are called, are all spun from Sea Islands, which belongs to this class. When we are told that a single pound of this cotton is often spun into a thread about 160 miles long we can see that it must be exceedingly good and strong cotton to do this.

FIG. 4.—The Gossypium Barbadense.

Lint is the name given to the cotton which remains when separated from the seeds. Every other American type of cotton gives a greater percentage of lint than the Sea Islands cotton, though it should be stated that the price per pound is greater than any other kind of cotton grown in the States. There are from six to nine seeds in each capsule and the prevailing colour is black. A cotton grown in Egypt and known by the name *Gallini* is of the Sea Islands type and has been produced from seed of the G. Barbadense. It should be added that the colour of the flower is yellow and that in India this plant is known by the name of Bourbon Cotton.

Gossypium Herbaceum.—As indicated by the name, this type is herbaceous in character, especially the cultivated type. When Lamarck classified this tree, he gave it the name Indicum because he considered most of the Indian types and some of the Chinese belonged to this particular species. India, too, is considered by Parlatore to have been the original home of the herbaceous type, and he specially fixes the Coromandel Coast as the first centre from which it sprang. There is much conflict of opinion in localising the primitive habitat of this type, and it is now thought that the present stock is probably the result of hybridisation of several species more or less related to each other. However, the areas in which this class of cotton grows are very numerous and extensive, for we find it growing in India, China, Arabia, Persia, Asia Minor, and Africa. A very characteristic feature of this plant is that it quickly decays after podding, when cultivated as an annual.

The *Vine Cotton* grown in Cuba belongs to the herbaceous type and is remarkable for its large pods, which contain an abnormal number of seeds.

The so-called "Nankeen" cottons are said to be "Colour variations" of the herbaceous Cotton plant. Many varieties of Egyptian cottons are produced from this particular class, as well as the Surat cotton of India.

A feature which distinguishes this type is that the seeds are covered with two kinds of fibre, a long and short, the latter being very dense. The process of taking the longer fibre from the seed must be very carefully watched, as it becomes a troublesome matter to remove the shorter fibre when once it has come away from the seed with the longer. Hence great care should be taken in gathering this class of cotton. Another point which should not be lost sight of is, that the herbaceous type of Cotton plant readily hybridises with some other varieties and the result is a strain of much better quality.

Gossypium Hirsutum.—This variety is so called because of the hairy nature of every part of the plant, leaves, stems, branches, pods and seeds—all having short hairs upon them. By Dr. Royle it is considered a sub-variety of the Barbadense cotton, and by other American experts it is given as synonymous with G. Herbaceum. However this may be, the plant has certain well-defined characteristics which possibly entitle it to be considered as a distinct type. It has been asserted by a competent authority that the original habitat of this particular cotton was Mexico, and that from this country cultivators have imported it throughout the sub-tropical districts of the world.

It is also stated that longstapled Georgian Uplands cotton belongs to the Hirsutum variety. In fact most of the types cultivated in America fall into this class. Parlatore also considers it to be indigenous to Mexico, and states that all green seeded cotton which is so extensively cultivated has been obtained from this type originally. M. Deschamps, in describing the Hirsutum species, says it is divided into two varieties, one having green seeds, being of a hardier type, and the other having greyish seeds, being more delicate and growing in the more southern districts of the States.

Gossypium Arboreum.—This plant attains treelike proportions, hence the name Arboreum. In some cases it will grow as high as twenty feet. It is also known by the name G. Religiosum, because the cotton spun from this plant was used only for making threads which were woven into cloth for making turbans for the priests of India. Dr. Royle on one occasion while in that country was informed by the head gardener of a Botanical Garden at Saharunpore that this cotton was not used for making cloth for the lower garments at all, its use being restricted to turbans for their heads, as it was sacred to the gods. That is why it also received the name, "*Deo Cotton.*"

One or two interesting features of this type may be pointed out. The colour of the flowers is characteristic, being brownish and purply-red and having a dark spot purple in colour near the base of the corolla, this latter being bell-shaped. Like the herbaceous type two kinds of fibre are found on the seed and great care is needed in the separation of them. Also, it should be pointed out that the fibres, in this class are with difficulty removed from the seed, clinging very tenaciously to it.

The Arboreum type is indigenous to India and along the sea board washed by the Indian Ocean. The fibre from this species is much shorter in average length than any of the preceding varieties.

Gossypium Neglectum.—This too is an Indian cotton, and according to Royle the celebrated and beautiful Dacca cotton which gives the famous muslins, as well as the long cloth of Madras, are made from cotton obtained from the Neglectum variety. An important feature of this plant is the small pod which bears the fibre and the small number of seeds contained in each septa of the capsule, being only from five to eight in number. Like some preceding forms, the seeds carry cotton of two lengths, the shorter of the two being ashy green in colour. The longer fibre is harsh to the touch and white in colour. In many points it is very similar to the Arboreum type and is considered by some botanists to be a cross between the Arboreum species and some other. It does not attain any great height, being often in bush form under two feet. The country of Five Rivers or the Punjaub, North West Provinces and Bengal, are the districts in India in which it is mostly cultivated as a field crop. It has a high commercial value, forming the main bulk of the cotton produced in the Bengal presidency. This plant is indigenous to India.

Gossypium Peruvianum.—So called because Peru was considered to be the habitat of this cotton. By some authorities this particular species is for all practical purposes synonymous with the first type described—viz., Barbadense. By others it is said to be closely allied to the Acuminatum variety, so named because of the pointed character of its capsules and leaves. Perhaps the most striking feature of this plant is the colour of the seeds, which is black. Another interesting point about the seeds is that they adhere closely to each other, and form a cone-like mass. Brazil is the home of this particular species, though it is cultivated here in two forms, as "Tree Cotton" and as "Herbaceous Cotton." The former is also known by the name Crioulo or Maranhâo Cotton or short Mananams. It appears also that the Tree Cotton is one of the very few which does not suffer from the depredations of the cotton caterpillar. What is known as "Kidney Cotton" belongs to this species, which is sometimes named Braziliense. The name kidney is given because of the peculiar manner in which the seeds are arranged in the capsule, adhering together in each cell in the form of a kidney.

The most important countries in which it is grown are Brazil and Peru, though it is produced in other districts outside these countries, but not to any great extent.

A very curious cotton which receives the name of "Red Peruvian" is also produced in South America. On account of its colour, it has only a very limited sale. This is owing to the difficulty there is in blending or mixing it with any other cotton of similar quality.

Cottons known generally as Santos, Cæra and Pernams are not of this species—viz., Gossypium Peruvianum, but belong to the first and second of the types already described.

The Strength of Cotton Fibres.—Mr. O'Neill some years ago made many experiments with a view to obtaining the strengths of the different fibres, and the following table compiled by him, will be of interest to the general reader.

Sea Islands	83.9 mean breaking strain in grains
Queensland	147.6 ""
Egyptian	127.2 ""
Maranham	107.1 ""
Bengueld	100.6 ""
Pernambuco	140.2 ""
New Orleans	147.7 ""
Upland	104.5 ""
Surat (Dhollerah)	141.9 ""
Surat (Comptah)	163.7 ""

From this table it is arguable that the strength of fibre varies according to the diameter, that is to say, the fibre with the thickest diameter carries the highest strain. The order, therefore, in which the fibres would fall, according to strength, would be, Indian, American, Australian, Brazilian, Egyptian, and Sea Islands last.

The Chemistry of the Cotton Plant.—Messrs. M'Bryde & Beal, Chemists in the Experimental Station in Tennessee, say, "As a rule our staple agricultural plants have not received the thorough, systematic chemical investigation that their importance demands." It would appear that until recent times the above statement was only too true. Now, however, the United States Government and others have instituted experiments on a large scale, and everything is now being done in the direction of research, with a view to improving the quality of this important plant.

A complete Cotton plant consists of roots, stems, leaves, bolls, seed and lint. Now if these six parts of the plant be weighed, they vary very much, proving that some of them are more exhaustive than others, so far as the fertilizing matters found in the soil are concerned. For example, if water be discarded in the calculation, though this takes up a fair percentage of the total weight, about 10, it is found that the roots take up by weight over 8 per cent. of the whole plant, stems over 23 per cent., leaves over 20, bolls over 14, seeds over 23, and lint only 10½ per cent.

Now this statement is interesting as showing one or two important features. The weight of the seed is seen to be nearly a quarter of the whole plant, while the stems and leaves together take up nearly one half. A very small proportion by weight of the plant is taken by the lint.

A chemical analysis of the mature Cotton plant yielded the following substances:—

Water.   Potash.

Ash.   Lime.

Nitrogen.   Magnesia.

Phosphoric acid. Sulphuric acid.

Insoluble matter.

Of ten analyses made with the cotton lint (which takes up about 10½ per cent. of the whole) M'Bryde states that the average amount of water found was 6.77, ash 1.8, nitrogen .2, phosphoric acid .05, potash .85, lime .15, and magnesia .16.

He very pertinently remarks also "that if the lint were the only part of the plant removed from the land on which it is grown, cotton would be one of the least exhaustive of farm crops. The only other part which need be permanently lost to the soil is the oil, which also contains very small amounts of fertilising constituents." In connection with this he further says "that even when the seed is taken away along with the lint, cotton still removes smaller amounts of fertilising materials from the soil than either oats or corn." It should be borne in mind that the soil upon which cotton is cultivated lies fallow for a greater part of the year, and the fact of absence of cultivation, with consequent non-fertilising and non-enriching of land, must tend in the direction of soil exhaustion by the Cotton plant.

Another useful and important fact in connection with the Cotton plant is the medicinal use to which the roots are put. According to the *American Journal of Pharmacy*, the bark from the roots of the Cotton plant contain an active ingredient which in its effects is very much like ergot.

Chemical investigations have conclusively proved that the ripe fibre of the Cotton plant is composed of the following substances:—

Carbon, Hydrogen, Oxygen, and they tell us that when cotton is fully ripe it is almost pure cellulose.

Dr. Bowman has pointed out that the percentage of water in cotton fibre "varies with different seasons from 1 to 4 per cent. in the new crop, and rather less as the season advances. Above 2 per cent. of moisture, however, seems to be an excessive quantity even in a new crop cotton, and when more than this is present it is either the result of a wet season and the cotton has been packed before drying, or else it has been artificially added."

About one fifth of the whole plant by weight consists of the seed, and an analysis of this shows them to be composed of water, ash, nitrogen, phosphoric acid, potash, soda, lime, magnesia, sulphuric acid, ferric oxide, chlorine, and insoluble matter.

As a commercial product seeds are exceedingly valuable, and yield the following substances:—oil, meal, hulls, and linters. When the hulls are ground they receive the name of cotton seed bran. The inside of the seed, when the hull has been removed, is often called the kernel and is sometimes also designated peeled seed, hulled seed, and meats. It is this kernel seed which, when properly treated, yields large supplies of oil and meal.

CHAPTER II.

COTTON-PLANT DISEASES AND PESTS.

There are several classes of agents all of which act injuriously more or less on the Cotton plant.

1. Climatic changes, including hygrometric variations of the atmosphere, and extremes of heat and cold.

2. Insect pests.

3. Physiological diseases of the plant.

4. Blights caused by fungi.

It has been pointed out in the early pages of this story, how very sensible to changes of heat and cold, the Cotton plant is, especially in the early growing period. When the plant has just risen above the ground, and is beginning to spread its roots, too great an amount of heat would be fatal to its further growth.

Instances could be given where very serious decreases in the production of cotton in the States especially have taken place, due entirely to unusually high temperatures which obtained during the early growing period of the Cotton plant.

Extremes of frost are likewise fatal to the growth of the young plant. By the beginning of April, frosts have as a rule disappeared, and no further fear need be felt on that account, though if the end of winter has been abnormally warm, and the young plants have been making leaf too quickly, it will be readily seen how fatal a sharp frost or two must be to the young and tender plant. There are cases, however, when a frost is beneficial.

Then again, while rain is needed in fair quantity, too much of it is followed by rot and myriads of pests. If the planter desires anything at all when his crop is ripe, it is fine weather in which to gather his harvest.

Frequently large quantities of cotton are left on the plantations, because it is too wet to gather it. This happened a few years ago to an unusual extent, when a vast quantity of cotton had to be left upon the fields.

Of all the injurious agents most dreaded in the cotton-growing districts of the globe, none are so widely spread or so disastrous as "insect pests."

They attack different parts of the plant during its growth, and when the bolls are formed they commit great havoc among these by boring through and completely ruining the immature fibre. Then again, while the plant is young, they may attack the most tender portion of the plant, viz., the new and young leaves found at or near the top. This they soon clear and make their way as caterpillars down the plant, and they frequently clear it as though the leaves had been plucked off.

So completely do they do their work that it has been calculated in certain years the loss on this account alone cannot have been far short in America of 3½ million pounds in one year.

Of the chief forms of insect pests, two specially stand out into prominence, both of which belong to the moth tribe of insects, viz., *Alethia argillacea* or Cotton Caterpillar, and the *Heliothis armiger* or Cotton Boll-Caterpillar.

The operations of the former are mostly confined to devastating the leaves and buds, while the latter confines its special attention to the bolls which, were they allowed to ripen, would burst with cotton.

The eggs of the former, too, are laid on the under side of the upper leaves and vast numbers are deposited. The moth flies by night, and the eggs laid are extremely difficult to discover—indeed it takes an expert to quickly find them.

Usually, about midsummer, the eggs are hatched in three or four days and then comes the period for spoliation.

All that is tender is assimilated, usually the under side of the young tender leaves found at the top of the plant.

During this stage of its existence the caterpillar moults five times and the larva period varies somewhat according to the weather from one to three weeks.

The chrysalis or pupa state covers from one week to four, and at last emerges as a beautiful olive gray moth with a purplish lustre.

In about four days the female commences to lay eggs very rapidly and will lay sometimes as many as six hundred during its life. No wonder, then, with several generations during a season and vast numbers of moths, that untold damages can be wrought by these particular insects in a single season.

A number of remedies has been successfully applied in the direction of spraying various chemical solutions, and in sowing plants which have had the direct effect of reducing the spread of this terrible pest. Its method of working can be seen on referring to Fig. 4.

Now the Boll-Caterpillar, though it lives much in the same way as the Alethia, has a very different method of procedure so far as its destructive habits are concerned.

And its fields and pastures, too, are by no means confined to one continent, or to one kind of plant, for it attacks both the tomato and corn plants. According to Dr. Howard, "It feeds upon peas, beans, tobacco, pumpkin, squash, okra, and a number of garden flowering plants, such as cultivated geranium, gladiolus, mignonette, as well as a number of wild plants." As the name indicates, the Boll-Caterpillar makes the boll its happy hunting-ground. The eggs are laid in the same way by the parent moth as the Cotton Caterpillar or Alethia, and when hatched the young powerfully jawed caterpillar makes its way to the newly-formed boll, and applying itself vigorously, very soon gains an entrance. Here it rests for a time, eating away at the best it can find. It ultimately emerges and is transformed into the pupa, taking up its quarters in the ground, until the next change takes place, when in a week or two's time it appears as a moth much the same in size as its cousin the Alethia, but coloured ochre yellow to dull olive-green and being more varied in its markings. It will lay during one season about 500 eggs.

Many remedies have been applied for the extirpation of this particular insect, but these only seem to have met with partial success. It will readily be seen how much more difficult this pest is to deal with than the preceding one. Living as it does in the boll and in the ground for a great part of its existence, it will be exceedingly difficult to get at.

In Mexico what is known as the Cotton-Boll Weevil (*Anthonomus grundis*) appears to do great mischief to the Cotton plant. It does most damage during the larvæ stage, eating up the tender portions of the boll while in residence here. When matured it is only a little under half an inch in length.

Many other insects act injuriously upon the Cotton plant, but the following may be taken as the chief: Cotton Cutworm (*Feltia malefida*); Cotton lice (*Aphis gossypii*). Among the lepidoptera may be mentioned, *Cocæcia rosaceana*, or "Leaf-roller," so called from its habit of curiously rolling the leaves of the Cotton plant and then feeding inside the roll. Then grasshoppers and locusts occasionally do some damage, as well as a beetle named *Ataxia crypta*, which is noted for attacking the stalks of the Cotton plants, but it should be pointed out this beetle does not prey upon healthy and vigorous plants at all.

Scores of other insects could be mentioned as injurious, though some of them do but very slight damage indeed to the Cotton plant.

It does appear, however, from long years of experiment and observations, that little damage needs to be feared if the plants, while growing, and during the formation of the boll, can be carefully watched and guarded. The plants when matured are better able to withstand the onslaughts which these predaceous insects make upon them.

Then again, there are large numbers of physiological diseases of the cotton due to inherent weakness of the plant or failure of assimilative processes.

And lastly, vast numbers of fungi, too numerous to mention here, work serious injury to leaf, flower and boll in certain seasons of the year.

CHAPTER III.

CULTIVATION OF THE COTTON PLANT IN DIFFERENT COUNTRIES.

From what has already been said, it will be quite clear that the Cotton plant will only successfully thrive in those regions on the earth's surface where there are suitable temperature and soil, and a proper and adequate supply of moisture both in the atmosphere and soil. When the 45th parallel of North Latitude is reached, the plant ceases to grow except under glass or in exceptionally well favoured and temperate districts. Below the Equator the southern limit is the 35th parallel.

With a model of the globe before him, the reader will see, if he mark the two lines already named, what a small belt the "Cotton-growing zone" is, compared with the rest of the globe, and yet in 1901 it is estimated that no fewer than 10,486,000 bales of 500 lbs. net average each were produced in the United States alone, 695,000 came from the cotton fields of India, from Egypt 1,224,000, an increase of 600,000 bales in ten years. This vast quantity does not include what was produced in other countries, which we know in the aggregate was very considerable.

American Cultivation of the Cotton Plant.—Perhaps no country illustrates the fact so well as does the United States, that the variations in the quality of cotton are very largely—it may be said almost entirely—due to distance from sea board, height above sea level and difference of soil.

The surface geology of the Southern United States as a whole, is of a most diversified character, and the following States in which cotton is produced, in many cases show a similar variation.

North Carolina.   South Carolina.

Georgia.   Florida.

Arkansas.   Tennessee.

Alabama.   Mississippi.

Louisiana.   Texas.

Perhaps Texas shows the greatest number of distinct soil areas, viz., eight. Height above the sea level has also a considerable influence upon the plants cultivated, and only the hardier and more robust types are to be found on the more elevated lands. At the beginning of the nineteenth century South Carolina produced more cotton than any other State. Fifty years later, Alabama was to the front. Ten years later, Mississippi led the way, and in 1901 Texas occupied the premier position with 3,526,649 bales, followed in order by Georgia and Mississippi.

The following table from Bulletin 100 of the Bureau of the Census, Department of Commerce and Labor, gives the acreage devoted to cultivation of cotton in 1908 as follows:

Alabama	3,591,000	acres.
Arkansas	2,296,000	"
Florida	265,000	"
Georgia	4,848,000	"
Louisiana	1,550,000	"
Mississippi	3,395,000	"
Missouri	87,000	"
North Carolina	1,458,000	"
Oklahoma	2,311,000	"
South Carolina	2,545,000	"
Tennessee	754,000	"
Texas	9,316,000	"
Virginia	28,000	"
	32,444,000	"

The figure for Missouri includes other cotton-producing localities not named.

Before dealing with the actual cultivation of cotton, as carried on in the States, it will be well to briefly name the kind of soils which are met with in this cotton area. Generally speaking, soils are divided into the following classes:—

Clayey soils.
Clayey loam soils.
Loamy soils.
Sandy loam soils.
Sandy soils.

This classification is determined by the relative percentage of sand and clay.

In the States we have all these types, and in some districts they lie within easy reach of each other. It should be pointed out that sufficient and uniform heat and humidity are essential to the production of good cotton crops, and as the sandy soils are of an open character, it is plain that moisture will readily pass from these, while the heavy clays act just in the opposite direction, viz., prevent the uniform evaporation of the moisture within them; hence, as a rule, clayey lands are moist and damp, and it has been found from observation that on lands of this class, a good deal of wood and leaf are produced, and but little fruit relatively. A matter therefore which must not be lost sight of, is that a suitable texture should be found, or, in other words, the amount of sand and clay in the soils should be in the right proportion. Of course, however suitable a soil may be, if the climatic conditions are adverse, only failure can result. Given good land, properly drained and a suitable temperature, together with an uniform supply of moisture, heavy crops may be expected. Sudden changes in the temperature, and variations in the amount of moisture, certainly act deleteriously upon the plant, especially during the period in which the young one is growing. There is a great difference between a wet soil and a moist one, and there is perhaps nothing so much dreaded by the planters as a sodden soil. Up to the end of July the soil should be continuously and uniformly moist, and it would appear that, provided this condition is satisfied, there is every likelihood of a heavy crop resulting, if the temperature has been anything like suitable. Looked at from every point of view, therefore, the best and safest soil in which to grow cotton is a deep loam where there is every probability of the necessary conditions being fulfilled.

As compared with sixty years ago the present methods of cultivation show very great differences. Most of us are acquainted with the conditions of labour which existed at that time. Mrs. H. Beecher Stowe, in her pathetic and life-like story, "Uncle Tom's Cabin," has given us such a glimpse into slave life that she has placed us all under lasting obligations to her. Happily all that has gone and the slave, as such, is now known no more in America. Three causes are said to have done more to change the methods of American cotton cultivation than anything else, viz.:—

The Civil War.
The abolition of slavery.
Introduction of artificial fertilisers.

There are those who affirm to-day that the last-named has been the most potent factor of the three.

In many cases, previous to the war, crop after crop was grown upon the same land without any thought of returning those elements, in the form of manure, to the earth, which it so much required. But immediately after the conclusion of the war, the conditions of labour were changed and it became a matter of absolute necessity to find something which would give life to the land, hence the introduction of fertilisers. It is stated on the authority of Dr. White of Georgia, that it would be "difficult to conceive how cotton culture could have been continued or sustained but for the use of such manures."

In a work of this kind it is impossible to describe in detail the various methods of cultivation adopted in the several cotton States, but the following will give a fair idea of what actually takes place on a large cotton plantation, assuming that the land is well drained. It should be said here draining has not received that attention which it ought to have done, and many of the failures put down to other causes are now known to have been due entirely to bad drainage. As an alternative to proper drainage, the practice of raising the Cotton plant beds and cultivating them to greater depth, has been followed. Most of the planters are too poor to drain properly, and so adopt the banking method, though in the long run this is the more expensive of the two.

Let us assume that the cotton crop has all been gathered. We have an immense quantity of old cotton stalks which need removing. This is usually done before February. As a rule, the litter is gathered into heaps and burned. Ploughing and harrowing next follow, and ridges are formed which in the elevated districts are not quite so far apart as in the low-lying areas. We can see that in the latter districts the plants will be much more prolific and grow to a better state of perfection, hence more room must be allowed for them. These ridges lie, in some cases, 3 feet apart and in others 4 and 5.

Especially when manures or fertilisers have been used, bedding up is generally adopted.

As is to be expected in a country like America, the very best and most approved methods of cultivation are followed, hence the old system of sowing seed by hand is discarded, and seed-planting machines are now coming into general use. The distance apart which the seeds (about five or six in one hole) should be set, is still a moot question, but it is generally admitted to be unsafe to plant at greater distances than 12 inches. When sown, a light covering is put over, and in a few days—about twelve generally—the tiny plants make their appearance. Two or three days after, another leaf is seen, and it may be said that the real and anxious work of the cultivator now begins. In the Carolina districts this will happen about the end of April. The planting in the more southern States will take place earlier. What has next to be done is very particular work, viz., cutting down and thinning the plants, which, if allowed to grow, would simply choke one another. Here and there at suitable distances, groups of plants in the same row are selected as "stands" or groups of plants from which will be selected the best plant, which is allowed to go forward in its growth; all the rest being chopped out or weeded out.

Banking up or bedding up is the next process, and this is done running the plough in the spaces between the ridges or practically over the old cotton bed of the preceding season. This will improve the ventilating power of the bed considerably and prevent somewhat the logging of the soil, which is extremely undesirable. The plough is immediately followed by the field labourers, whose work is now to draw the loose soil round the Cotton plants. This last process of "hauling" completes the labourers' work for a time, and is done for the purpose of keeping the plant erect and preventing it from falling down. This hauling process is repeated until July, when only one plant is left out of the five or six which were planted originally. After four haulings, which are completed as a rule by the end of July, the productive processes may be said to be completed. If the weather has been favourable and the soil kept fairly moist, a good crop may be fully anticipated. What the planters like to see during the growing period is a summer in which the sun shines every day, accompanied by those frequent and gentle showers which clean the plant and give the necessary humidity to the atmosphere and soil. Two things are dreaded by the planter—excessive heats and abnormal showers. The bloom appears about the middle of June and a couple of months after this the plants are ready for picking. This operation usually is carried on from the beginning of September or end of August right on into November, sometimes through this month into December. Here are given a few particulars which have been collected by Shepperson bearing on this particular subject.

STATES.	Usual date to begin Preparing the Land.	Usual date to begin Planting.	Usual date to finish Planting.	Usual date to begin Picking.	Usual date to finish Picking.

N. Carolina	Feb. 25	April 15	May 10	Sep. 1	Dec. 10
S. Carolina	Mar. 5	April 15	May 7	Aug. 15 to Sep. 1	Dec. 1
Georgia	Feb. 1	April 10	May 1	Aug. 15 to 20	Dec. 1
Florida	Jan. 20	April 1	May 1	Aug. 10	Dec. 1
Alabama	Feb. 1	April 5	May 10	Aug. 10 to 20	Dec. 15
Mississippi	Feb. 1	April 5	May 10	Aug. 10 to 20	Dec. 15
Louisiana	Feb. 1	April 1	May 10	Aug. 1 to 15	Dec. 15
Texas	Jan. 15	March 15	May 10	Aug. 1	Dec. 20
Arkansas	Feb. 15	April 15	May 15	Aug. 15 to 20	Jan. 15
Tennessee	Mar. 1	April 15	May 15	Sep. 1 to 10	Jan. 15

Other Cotton-producing Countries in America.—In addition to the States, which have already been named, there are other cotton-producing countries in the Western Hemisphere, among which are the following:—

Brazil.   Mexico.

West Indies. Peru and the South Sea Islands.

Cultivation of Cotton in Brazil.—From a very remote period, cotton has been cultivated in Brazil. Early in the sixteenth century historians refer to the uses to which cotton was put at that time. Seguro, in his work describing the customs of the ancient people who lived in the Amazon valleys, says that the arrows used in connection with their blowguns were covered with cotton. It is probable that, before the dawn of the eighteenth century, the cultivation of cotton was practised more or less throughout the country. Up to thirty years ago, it looked as though the cotton-growing industry in Brazil was likely to be an increasing and profitable business. Owing, however, to many causes, the trade has not grown as was to have been expected.

Among the chief of these causes are:—

1. Laxity of method in cultivating.
2. Poor means of transmission.
3. Severe competition by the United States.
4. Disturbed condition of the country.

All these have helped to keep down an industry which at one time bade fair to be a source of great income to the country.

Tree Cotton and Herbaceous Cotton are both cultivated in Brazil. The best kinds of Sea Islands have been tried, but have not succeeded.

Compared with the United States, the methods of cultivation pursued in Brazil are exceedingly primitive and irregular. No such thing as ploughing or preparing of the soil is adopted.

The only preparation seems to be to rid the land of cotton stumps, and this is done in a somewhat careless and indifferent manner. It would seem that as little labour as possible is expended upon the land in preparing it for the reception of seed. Hilaire's aphorism—"Nothing in this country is less expensive, or more productive, than cotton culture"—would seem, when the facts of the whole case are known, to be perfectly warranted so far as Brazil is concerned. Certainly, from a climatic point of view, this country is exceptionally well favoured, an equable and suitable temperature together with an adequate supply of earth and air, moisture and rich alluvial soils, a long dry season for picking extending over many weeks—all point to an ideal cotton-growing area. In fact, there is no reason why a crop of at least 40,000,000 bales should not be obtained annually in Brazil, if needed. At present, only about one three-hundredth part of this is grown. The cotton-growing centres are Minas Geraes, Bahia, Fernando de Noronha, Rio Janeiro, Sao Paulo.

Cotton Cultivation in Mexico.—The cultivation of cotton has for many centuries been carried on in Mexico. Much the same drawbacks exist here as in Brazil, viz., lack of labour, poor railway system, high rates for transmission, and indifferent methods employed in cultivating.

Mexico enjoys a splendid geographical position and would prove, if the business-like habits and methods obtained as in case of the States, one of the most serious competitors of its adjacent Northern neighbour.

The best cotton is produced in the State of Guerrero on the Eastern side, though the greater part—about one half—of the Mexican crop is grown in Laguna district, which lies in the Coahuila country. There are three distinct areas of production in Mexico, viz., along the Eastern coast, along the

Western coast, and on the Central tableland. In the Western area irrigation is resorted to.

In the year 1898, 100,000,000 pounds of cotton were grown, though all or nearly all of it was used at home. Within the last twenty years many mills have been erected in this country, and this will account for the large quantity of cotton consumed at home. The poorest Mexican cotton is produced in Chiapas. Acapulco, near the mouth of the Grande del Norte River, is the chief Mexican cotton port on the Eastern coast.

Cotton-growing in Peru.—It would be a difficult matter to fix a time when cotton was first grown in Peru. Pizarro, who conquered this country early in the sixteenth century, found that the natives were fully engaged in the growing and spinning of cotton. Dr. Dabney, Assistant Secretary of the U.S.A. Agricultural Department, states that he has seen a cloth made of cotton recently taken from one of the Peruvian mummies which must be many, many centuries old. There is not the slightest doubt that the Cotton plant is indigenous to Peru.

Thirty-five years ago Liverpool received no less than 300,000 pounds weight of cotton from Peru, and three years later over 4,000,000 pounds. During the last decade of the century it exceeded 6,000,000 pounds to England alone. Two kinds of Peruvian cotton are grown—smooth and rough. This latter is a rough, strong fibre, and is exceptionally well adapted for mixing with wool in the manufacture of hosiery, and a greater part of this cotton coming in England is used in the hosiery trade. The plant from which it is produced is a perennial, and for six or seven years is said to give two crops a year. Owing to the peculiarly favourable climate of Peru and the suitability of the soil, it is exceedingly improbable that any strong competitor will come to divert the Peruvian trade, so that for some time yet we may look to this country supplying the hosiery trade with rough Peruvian cotton. The importations of Peruvian cotton into the United States for 1894-95 were 24,000 bales; for 1895-96, 24,603 bales; for 1896-97, 16,604 bales.

The Cultivation of Cotton in India.—There are other Asiatic cotton fields besides those of India, viz., China, Corea, Japan, the Levant, and Russia in Asia. The term "India" will be used in a somewhat restricted sense in this section, and will cover only that huge triangular-shaped peninsula lying to the south of Thibet in Asia. It is 1800 miles in width and nearly 2000 miles in length. The total area, not including Assam and Burmah, is about 1,300,000 square miles, the native states alone covering 595,000 square miles.

Out of the 28° of North Latitude through which India stretches, no less than 15½° are in the tropics, the remainder being in the Temperate Zone.

The climate, owing to a number of circumstances, such as different altitudes and uneven distribution of moisture, is exceedingly varied.

During the months April to September the sun, during the day or some part of it, is overhead. Consequently the heat received will be greater than over the ocean at the south, taking a similar area. A direct cause of this is the starting of winds which receive the name of monsoons. These blow from the S.W., and bring vast quantities of moisture with them. This moisture-laden wind is partially robbed of its load as it strikes the Western Ghats and consequently much moisture is deposited here, giving rise to many valuable rivers which water the Deccan or Central Tableland of India. The Mahanuddy, Godavari, Kristna, and Kauvari are rivers fed by the S.W. monsoon. Then, again, the low-lying lands near the mouth of the Indus, the great desert of Rajputana, the peninsula of Gujerat and the district of Malwa—all allow, by reason of their low-lying nature, the S.W. winds to pass over them laden as they are with vast quantities of moisture. They travel on till they meet the Himalayas, where again they help to swell the volume of the waters in the rivers Ganges and Indus. When the N.E. monsoons blow they do not carry anything like the amount of moisture which the S.W. monsoons do, as their areas of collection are very much more limited. Consequently this part of the year is usually a dry one (viz., from October to March).

Thus it will be seen that the great plain of Southern India is much less watered than the more Northerly portions and consequently is much less fertile. This fact must be borne in mind as the cotton-growing areas are described and indicated.

India, which grows more cotton than any other country in the world (the States excepted), may be said to possess four distinct areas for the production of commercial cotton. They are—

1. Central Tableland or Deccan.
2. Valley of the Ganges.
3. Western India.
4. Southern India.

and the above order shows them also according to their commercial importance.

Central District.—This is a vast plateau bounded on the north by the Vindhya mountains, on the east and west by the Ghats of those names, and on the south by the River Krishna. As is to be expected, the collecting and exporting of the cottons grown in this district are done at Bombay. The finest cottons grown in India are produced in this region.

Four centres stand out prominently in the production of cotton, viz., Dharwar, Hyderabad, Nagpore and Berar. The soils generally in the Deccan are very rich and capable of retaining moisture during the growing term of the plant's life. What are known as the black soils of India are to be found plentifully in this district, and these are exceedingly rich in mineral matter. Nagpore should specially be named, as it is in this province that the finest cotton grown in all India is produced, viz.:—

"Hingunghat Cotton."

"Oomrawattee Cotton" is the name given to a special kind which is produced in the province of Berar. It is sometimes called "Oomras." This district lies in the "Nizam's Dominions" and is watered by several tributaries of the Tapti and Godivari. It possesses a soil which for richness and fertility has no equal in India.

With the exception of Bengal, this district is more plentifully supplied with rivers than any other part of India.

FIG. 5.—An Indian cotton field.

The Dharwar district is noted for its cottons, for two or three reasons. It was in this region that in 1842 New Orleans cotton was planted with a view to its ultimately being cultivated here. As the climate and soil are very similar to some of the districts in the Mississippi valley, it succeeded beyond anticipation. Dharwar lies S. W. of the province of Hyderabad near the sea, and almost touches 15° N. Latitude.

The Valley of the Ganges District cannot be said to grow very good cotton, though it was in this region, at Dacca, that in former days the cotton which was afterward made into the celebrated Dacca muslin was grown.

By far the greater part of the fibre produced in this district comes from two centres: (1) Bundelkhand, which lies 79° E. Long., and 25° N. Latitude (this is very near to Allahabad), and (2) Doab. As was pointed out in describing the monsoons, these two centres suffer by reason of droughts, owing mainly to their geographical position. They are subject also to severe floods, which are certainly against successful cultivation of cotton. The entire crop of the North West Provinces may be said to come from the districts of Doab and Bundelkhand.

Western India District.—The three centres for the production of cotton in the west, may be said to be Peninsula of Guzerat, the Island of Cutch and the Delta district of the Indus named Sind. The whole of these provinces lies in what may be called a dry area, missing, as was shown, much of the S. W. monsoon, which ultimately finds its way across country to the Himalayas. Consequently there will be little rainfall in this area, Sind and Cutch not more than 10 inches, some parts of Guzerat having much more.

This has a very serious effect upon the quality of the cotton produced.

The Surat, Broach and Sind Cottons, all poor types, are all grown in this part of India.

Southern India District.—This lies in the southern part of the Residency of Madras, and east of the province of Travancore. The Nilgiris and Shevaroy Hills are found here, as are also the Cauvery and Vaigai Rivers. The cotton districts best known are Coimbatore and Tinnevelley, both of which are admirably situated and well watered. The Calicut of fame which gave rise to the name Calico is also in this district. Tinnevelley lies almost at the extreme south of India on the Gulf of Manaar opposite to Island of Ceylon. Its cotton is well known, but is of a poor type. As far back as 1847, experiments carried out under the superintendence of Dr. Wright proved that this district was very suitable for the cultivation of American cotton. A fact interesting as well as instructive is given by him to the effect that in the southern part of India the crops universally failed where grown from the native seed, while those grown from American seed realised very fair amounts—better even than were obtained when good crops were got after using Indian seed.

The methods of preparing, planting, and cultivating the Indian Plants are exceedingly antiquated. In but few districts are anything like modern methods practised. Advantage however is taken of the period just preceding the rain monsoon and this differs a little according to the district.

Thus in Bengal, Berar, and Broach, May and June are usually taken for scantily preparing the land, and in Madras and Dharwar, August and September. This consists of turning over the soil and burying the old Cotton plants of the previous season which have been allowed to rot. As no fertilisers are used, these roots and branches at best make a very poor substitute. Ploughing, hoeing and other agricultural operations are of the rudest types and oxen are used for almost everything in the way of heavy labour. Farm implements, gearing carts, etc., are all of a style and differ very little from those used centuries ago. The seeds are sown broadcast, and almost everything is done by hand.

The plantations as a rule are much smaller than those in America, running from 5 to 30 acres. On the larger plantations the cotton is cultivated mainly by paid labourers.

The following table, by Shepperson, shows the acreage devoted to cotton of the different states in India:—

Bombay and Sind	5,021,000 acres.
Punjaub	1,177,000 "
N. W. Provinces	1,424,000 "
Bengal	153,000 "
Rajputana	549,000 "
Central India	503,000 "
Berar	2,307,000 "
Central Provinces	616,000 "
Hyderabad (Nizam's)	2,308,000 "
Madras	1,655,000 "
Mysore } Assam } Burmah (Lower) } Burmah (Upper) }	230,000 "
Ajmere and Meywara	40,000 "

15,983,000   "

Bombay, Kurrachee, Calcutta, Madras, Tuticorin and Cocanada are the chief Indian cotton ports.

Cotton-growing in Russia in Asia.—Lying immediately north of Persia and Afghanistan and south of Khirghiz Steppes lies an immense area much of which is now being cultivated and most of it very fit for the production of cotton. The Sea of Ural has running into it two very large rivers, Amu Daria and the Syr Daria, and it is in the neighbourhood of these two rivers where we find by far the greatest weight of cotton of Turkestan produced.

There are four important areas, viz., Syr Daria, the centre of which is Tashkend; Fergana, which lies between Samarcand and Bokhara; the district of Samarcand itself; and Merv, which stands on the Overland Railway. It appears that many attempts were made to introduce cottons of various types into this locality, but most of the delicate species failed. The Upland of America, however, survived, and has continued to succeed, thanks to the valuable help which the Government gave in the way of instruction and distribution of free seed.

The first Government cotton plantation was commenced at Tashkend, one of the termini of the Transcaspian Railway. Eight years ago there were upwards of a quarter of a million acres devoted to cotton cultivation.

During the American War (that period which quickened all the cotton-growing centres of the Eastern Hemisphere) the production of fibre may be said to have commenced in earnest in Turkestan, and so late ago as 1890 no less than forty-five and a half million pounds of good fibre were grown. Tashkend, it would appear, promises to hold its own, as it is determined to practise the best and most scientific methods in the growth of cotton; in fact, in very few centres outside this district, old and out of date operations are followed. Even in the districts of Fergana and Samarcand, the old wooden plough called a "sokha" is still in use.

Seed, as in the case of India, is mostly sown broadcast, and very little preparing of the land is done. Yet, in spite of these deficiencies, good crops are raised in many districts, capital soil and a most equable climate making up for the shortcomings of the planter. The formation of the Transcaspian Railway cannot but have an important influence upon the cotton-growing industry in Turkestan, running as it does through the very heart of the best land in the country. It should be noted that Bohkara annually produces over 50,000,000 pounds of cotton of the herbaceous type, and Khiva, another district lying still further east of those already mentioned, over 20,000,000 pounds.

Lying between the Caspian Sea and Black Sea, lies another district named Transcaucasia, which yields large supplies of cotton. It has 100,000 acres devoted to cotton, giving over 20,000,000 pounds per annum. North of Kokan, on the river Syr Daria, is a rising cotton district named Khojend, where annually 3,000,000 pounds of cotton of the American type are raised.

When we consider that the quantity of cotton carried by the Transcaspian Railway since 1888 has more than quadrupled, and that in ten years the quantity shipped has been increased from quarter of a million pounds to over 72,000,000 pounds, we can quite appreciate the significance of the statement that before long Russia will be able to grow all her own cotton for the medium and lower numbers of yarns.

Cotton-growing in China, Corea and Japan.—Japan, the land of the chrysanthemum, for many years now has been developing cotton-growing as well as cotton manufacturing. From evidence which the cold type of the Board of Trade gives, Japan bids fair to largely increase her trade with India to the disadvantage of the present suppliers.

Cotton-growing has been practised for some centuries in Japan, but it was not until the seventeenth century that anything like progress could be reported. From that time to the present the growth has been gradually on the increase.

Japan proper consists of the Islands of Niphon, Kiusiu, Shikoku, Yesso, and an immense number of smaller islands. Cotton cultivation is carried on mainly on the first three islands named, and in the following districts:—San Indo, Wakayama, Osaka, Kuantoebene, Hitachi and Suo.

Taken as a whole, the cotton grown in the best areas is good, though much of an inferior kind is produced. The most southerly area of Wakayama in Niphon yields the best cotton of Japan.

The length of the fibre generally is much less than the herbaceous kind. About 10 per cent. of the entire arable land is now under cultivation for cotton. As a rule, methods and processes are of a primitive kind.

Cotton-growing in Corea.—Lying directly to the west of Japan, this vast peninsula has of late years been developing its cotton-growing. Five centuries ago cotton was imported from China, and one sees on every hand the influence of the Celestials. The cultivated plant is of the perennial type, though it is planted annually, the old plants being dug up and burned, the ash being used as a fertiliser. Statistics at present are not to be relied upon, though it is supposed that something like three quarters of a million acres are now under cultivation, giving on the average about 250 pounds of cotton lint. As in the case of Japan very little of this is exported, all of it or nearly so being spun and woven at home on the most primitive of machines.

The chief districts engaged in growing cotton, nearly all of which lie in the southern portion of the peninsula, are Hwang-Hi, Kyeng-Sang, Chel-La, Kyeng Kwi, and Chung Cheog.

Cotton-growing in China.—Owing to the great difficulty of obtaining any reliable statistical information, it is impossible to give anything approaching accuracy as to number of pounds of cotton produced annually, or number of acres devoted to the cultivation of the Cotton plant. This much, however, is known, that for many centuries cotton cultivating has been followed and that there has been within recent years a great increase in the weight of the cotton crop as well as in the acreage. The type of plant most generally cultivated is the herbaceous, and the cotton resulting is only poor in quality. Little or no preparation is made before sowing seed, which is generally done broadcast. As a result there is much overcrowding, and as is inevitable, there is produced a stubby plant with small bolls and much unripe cotton. On the terraces of the hillsides something approaching cultivation is pursued, with the result of a better crop.

Usually twenty weeks intervene between planting and picking, this latter operation being mostly the work of children and women. The old cotton stalks are afterward collected and dried for fuel.

Very few large plantations exist in China, most of them being only a few acres in extent.

But little of the cotton grown at home is exported, most of it being spun and woven by women, though some of the fibre is sent to Japan.

Cultivation of Cotton in Egypt.—It is now over thirty years since Sir Samuel Baker, the great African traveller, wrote these words: "The Nile might be so controlled that the enormous volume of water that now rushes uselessly into the Mediterranean might be led through the deserts, to transform them into cotton fields that would render England independent of America."

The crop for the season 1900-01 was no less than 1,224,000 bales of 500 pounds each. Ten years ago only 868,000 acres were devoted to cotton cultivation as against 1,350,000 acres laid down to-day. Everything, then, points to Sir Samuel Baker's statement becoming an actual fact much sooner than the famous traveller himself anticipated.

Egypt enjoys many advantages over her competitors across the Atlantic. In the first place, she can get almost twice as much cotton from the acre, so productive is the soil. Labour is cheaper, and the plant itself when young is not subject to the devastating frosts so often met with in America.

Egypt is divided into three great areas:—Lower Egypt, which includes the whole of the Delta of the Nile; Upper Egypt; and Nubia. It is in the first-named district where the whole of Egyptian cotton is produced. At the present time immense sums are being spent on irrigation and drainage works, and as these are extended the areas devoted to cotton production will greatly increase.

At the present time five distinct varieties of cotton are cultivated—

Mitafifi.   Bamia.

Abbasi.   Gallini.

    Ashmouni-Hamouli.

The latter variety was originally known by a different name, Mako Jumel. For a long time Ashmouni cotton was the principal fibre exported, but Mitafifi is now in the front of all the other Egyptian cottons. A noteworthy fact in connection with Ashmouni is, that its cultivation is on the decline.

Sea Islands Gallini—as it was sometimes called—has practically ceased to be cultivated. Of Mitafifi and Bamia fibres, Mr. Handy, U. S. A., says: "The Mitafifi was discovered by a Greek merchant in the village of that name. The seed has a bluish tuft at the extremity, which attracted the merchant's attention, and on planting it he found that it possessed decided advantage over the old Ashmouni. It is more hardy, and yields a greater proportion of lint to the seed. At first from 315 pounds of seed cotton, 112 pounds of lint was secured, and sometimes even more. It is now somewhat deteriorated, and rarely yields so much, averaging about 106 pounds of lint to 315 of seed cotton. The Mitafifi is a richer and darker brown than the Ashmouni. The fibre is long, very strong, and fine to the touch, and is in great demand. In fact, it controls the market.

"Next to Mitafifi, Bamia is perhaps the most extensively cultivated variety in Lower Egypt. It was discovered by a Copt in 1873. The plant is of large size and course growth. It is later and less hardy than Mitafifi, and the fibre is poor as compared with that of Mitafifi and Abbasi, light and brown in colour, and not very strong. In general, it may be said that this variety is inferior to Mitafifi in yield, hardiness and length and strength of fibre."

Other places where Cotton is grown.—In Africa, on the eastern and western coasts, large quantities of cotton are produced. The following countries are specially suitable to the growth of cotton: Soudan, Senegambia, Congo River, Free States, and Liberia. Possibly, when these districts are more opened up to outside trade, and European capital and labour are expended, abundant supplies of cotton fibre will be given.

Cotton is also grown in the East Indies, at Java, Sumatra, and Malay States.

In the West Indies formerly, large supplies were yielded, but owing to the cultivation of other crops that of cotton has steadily declined.

Greece and Turkey both yield cotton which goes by the name of Levant Cotton.

CHAPTER IV.

THE MICROSCOPE AND COTTON FIBRE.

This story would be very incomplete if some reference were not made to the wonderful assistance which has been given to the study of cotton fibre by the microscope. As seen by its help, some striking peculiarities at once make themselves apparent. It is proposed, briefly, in this chapter, to do three things:

1. To describe the construction of a suitable instrument sufficient for a complete examination of fibres in general.

2. To indicate the chief microscopic features of cotton fibres.

3. To show how to exactly measure the lengths and diameters of fibres by means of micrometers.

First, as to the instrument: a good substantial stand is desirable, one that will not readily vibrate. The microscope shown in Fig. 6 is a cheap and commendable form, and good work can be done by this instrument, which is made by Ross, London. The stand carries the body-tube, and at the lower end is placed the objective, so called, because the image of the object (which rests upon the stage as shown) under examination is first focussed by it and conveyed along the body-tube.

The top end of the said tube contains the eye-piece, so named because by its aid the eye is allowed to receive the image duly focussed and enlarged.

As a rule, beginners work with one objective only, generally a one inch.

FIG. 6.—Microscope in position for drawing objects.

A much higher power than this is necessary if the fibre in question is to be seen at its best, and for the purpose of this chapter a quarter inch objective will be used.

Underneath the stage, which is pierced by a circular aperture, is a diaphragm. This regulates the quantity of light which is to be transmitted by means of the silvered reflector shown in the illustration.

As a rule, two reflectors are fixed in the same holder; one a concave mirror, the other a plane one. The former brings the rays of light to a point or focus while the latter simply passes the beam of light along just as it received it, viz., as a parallel beam of light.

In examining fibres the concave mirror will be of most use. An ordinary lamp is usually good enough for the light required, the one figured being very suitable and having a tube-like arrangement of wick. Behind the body-tube are two forms of adjustment, coarse and fine. The latter is worked by means of the milled screw, conical in shape, which is found immediately behind the coarse adjustment. The operator is supposed to have had some slight experience in the manipulation of the microscope. The slide is now placed upon the stage. Fine Sea Islands cotton is mounted in Canada Balsam and protected by a small circular cover glass.

Now rack down the body-tube by means of the coarse adjustment until within $1/16$ of an inch of the cover-glass of the slide. Now see that the light from the lamp is fully on the cotton strands. Rack up or down, as the case may be, with the fine adjustment, and a wonderful sight meets the eye, for the cotton viewed through the microscope is altogether unlike what we should expect it to be.

Running completely across the field are a number of strands, varying in thickness, form and natural twist. What is meant by natural twist is very clearly shown in Fig. 7.

Most people have seen india-rubber tubing or piping such as is used in the chemical laboratory or that often found attached to feeding bottles. Take about a foot of this and hold one end firmly. Abstract the air by means of the mouth, and it will be found that immediately the air is taken out the tube collapses. Now if the rubber be variable in thickness, here and there along these lines of least resistance will be found certain twists, and it is the same kind of twists which can be so distinctly seen as the cotton fibre is viewed through the microscope. They are exceedingly irregular in number, on equal lengths of the same single fibre. When they run for some length, and are fairly regular, the edges appear like wavy lines or corrugations. It will now be seen by the reader why these twists are so invaluable in spinning: locking and intertwining with each other, they materially assist the spinner in building up a long and continuous thread.

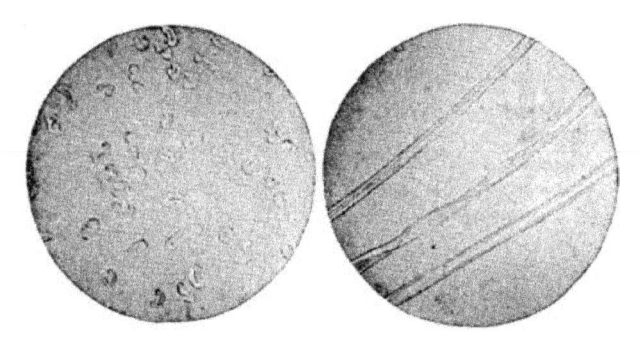

FIG. 7.—Transverse and longitudinal sections of cotton fibre.

Then, too, are to be seen lying close to the regularly twisted fibres a number of others which are very like ribbons, with here and there an apology for a twist, and further, a careful scrutiny will be rewarded by finding in what is reputedly the best cotton a number of filaments which do not display any twists whatever and are very much like the round tubing referred to a little while ago. Others again are quite flat, without any distinguishing twists whatever. These are said to be the half-ripe and unripe fibres, and give much trouble later on (if worked up with good cotton) to the dyer and spinner.

As the slide containing the cotton is moved laterally, it will be seen that this twisting of the fibre is continued for almost the whole length, and as many as 300 twists have been counted on a single filament. In some, the fibre tapers slightly, becoming more and more cylindrical as the end most remote from the seed is approached, until it is quite solid. These stiff ends soon disappear after the cotton has been treated in the early processes of manufacture. Thus there may be found in almost every sample of cotton what are called ripe, half-ripe and unripe cotton. The last-named kind result from—

1. Gathering the crop before the boll is properly ripened and matured.

2. Bad seasons; too much moisture and too little heat.

Then again in the same boll all fibres do not ripen together just as all apples on the same tree do not ripen together.

Immature or unripe cotton cannot be dyed, and when small white specks are seen in any dyed fabric they are often due to the fact that unripe cotton has been used in the manufacture of the cloth.

Measurement of the Cotton Fibre.—This is not at all a difficult matter, and the ordinary student may, by means of very simple and inexpensive apparatus, obtain fairly satisfactory results in the measurement of fibres.

There is a choice of one of three methods, viz.:—

1. By having the mechanical stage so arranged that the slightest displacement either to the left or right can be measured, and having the eye-piece so marked (generally a hair stretched across it) that when an object is to be measured, one side of it is made to coincide with this central line and the stage rack is worked left or right until the opposite side of the object is brought coincident with the central line again; the amount of displacement can then be readily obtained on referring to the graduated stage.

2. By having a stage micrometer and camera lucida.

3. By having two micrometers, a stage micrometer and eye-piece micrometer.

This latter method is certainly the least expensive, and for all practical purposes can be safely recommended.

A stage micrometer consists of a slip of glass $3'' \times 1''$ on which are marked divisions of an inch, usually $1/100$ths and $1/1000$ths. As a rule these markings are protected by means of a small cover-glass.

Eye-piece micrometers vary much in form, size and value, but the one which is here described is of the simplest type. It consists of two circular pieces of glass carefully cemented together. On one of the inner surfaces are marked usually the $1/100$ths divisions of an inch. In some $1/200$ths are marked. If the top lense of the eye-piece be unscrewed, a diaphragm will be found on which the eye-piece micrometer will easily rest. Screw on the top lense again and, generally, the eye-piece will be ready for use. If the micrometer is not properly in focus after a few trials, it may easily be made right. In order, then, to measure the diameter of a single fibre of Sea Islands cotton, fit in the quarter inch objective and place the stage micrometer in position on the stage. First, focus the fine lines which are plainly to be seen, and remember the lines which are farthest apart are $1/100$th of an inch; the others $1/1000$th of an inch.

As a rule, these lines run from N. to S. of the field; in other words, from top to bottom across the circles of light. Now look at the divisions in the eye-piece micrometer, which are $1/100$th of an inch apart.

It will be found often that an exact number of these divisions fill up one of the 1/100th divisions of the stage micrometer markings. If an exact number are not found, the draw-tube at the top end of the body-tube should be withdrawn until an exact number is found to lie within two lines of the lower micrometer.

Suppose twenty-two of the spaces on the eye-piece micrometer just cover one of the divisions (1/100th of an inch) on the stage micrometer. Then it is clear that each division of the former represents $1/100 \times 1/22$ of an inch, or 1/2200th of an inch. For every fresh objective used, a fresh estimation of eye-piece and stage micrometer ratio is necessary. Having now got in the eye-piece micrometer a unit of measurement, it becomes a comparatively easy matter to measure the fibre.

Remove the stage-micrometer and put a slide of Sea Islands cotton in its place. Focus the fibre and observe the number of divisions or parts of a division covered by any particular fibre, and its measurement is at once known. Thus if a single filament covers two of the divisions then it is 2/2200th of an inch in diameter, or 1/1100th of an inch. Exactly the same method is adopted if it is desired to measure the diameters of sections of the same fibres.

The making of the drawing of a fibre, either transverse or horizontal section, is not at all a difficult matter.

All that is needed is what is known as a camera lucida. This consists of a brass fixing for the eye-piece end of the body-tube and a small reflecting prism. This prism receives the image of the objective, and reflects it in this case at right angles downward on to a sheet of paper, which is placed beneath for the purpose of tracing the said image.

Focus the object, first having the microscope in a horizontal position. This will not be a difficult matter. Now remove the cap which fits on the eye-piece, and fix on the camera lucida as shown in the illustration (see Fig. 6). Adjust this until the image of the fibre is seen. Usually one or two smoke-coloured glasses are fixed below the prism, and these are now brought into position so as to allow the image of the fibre to pass through them. Place a sheet of drawing paper directly under the camera lucida, sitting as shown in the illustration. After a few trials it will not be a difficult matter to follow the outline of the image by means of a black lead on the paper as is shown in the figure. In this way many useful working drawings can be made, and a little careful calculation will give the amplification of the drawing after it is made.

CHAPTER V.

PLANTATION LIFE AND THE EARLY CLEANING PROCESSES.

After many months of anxious watching and waiting, towards the end of July or early in August, the planter may be seen to be constantly and wistfully looking for the appearance of the bursting bolls of cotton. Daily in the early mornings he is to be seen casting his eyes down the pod-laden rows of cotton plants, to see if he can count a few ripe open bolls as he stands at the head of a row. If this be so, he knows that his harvest is close at hand, and his pickers must be ready at any moment to begin what is certainly the most tedious and difficult work of the plantation, namely, picking the raw cotton from the bursting bolls.

While the planter has been on the lookout in the fields, necessary and important operations have been going on inside in the farm outbuildings. Sacks and baskets which can most expeditiously aid in the removal of the picked cotton from the field to the ginning factory are being got ready. To suit the young and old, tall and small, weak and strong, different sized bags and baskets are required, and after the marking and branding of the same, they are ready for being put into use.

Now the picking of cotton is not at all an easy operation, long continuous bending, a hot sun (for it is a rule scarcely ever broken that cotton must not be plucked unless the sun is shining upon it), a constantly increasing weight round the neck or on the arm, monotonous picking of the cotton from the bolls without bringing away any of the husk or leaf—all tend to make the work of the picker very trying and tiresome. The plantation hands must be early at work, and while the day is very young they are to be seen wending their way, ready to begin when the sun makes its appearance. Often the clothes of the workers are quite wet with the early morning dews. This is specially the case in September and October. By ten o'clock a hot blazing sun streams down upon the pickers as they diligently relieve the heavy-laden bushes of the white fleecy load of cotton. As each picker fills his or her bag, it is quickly emptied into a larger receptacle, and ultimately carried away to the gin house, where it is desirable the cotton should be housed before the night dews come on and consequently damage materially the cotton which the pickers have been careful to pick while the sun was on it.

Mr. Lyman, in his book on the Cotton Culture in the States, says: "It seems like very easy work to gather a material which shows itself in such abundance as fairly to whiten the field, but let the sceptic or the grumbler take a bag on his shoulders and start in between a couple of rows. He will find upon taking hold of the first boll that the fibres are quite firmly attached to the interior lining of the pod, and if he makes a quick snatch, thinking to gather the entire lock, he will only tear it in two, or leave considerable adhering to the pod. And yet he may notice that an experienced picker will gather the cotton and lay his fingers into the middle of the open pod with a certain expertness which only practice gives, the effect of which is to clear the whole pod with one movement of the hand."

Knowing how intensely monotonous and dreary the work of cotton picking is, Mr. Lyman advises the planters to allow a very fair amount of liberty so far as merrymaking is concerned, and he says on this point that "though too much talking and singing must interfere with labour, it is earnestly recommended to every cotton grower to take care to secure cheerfulness if not hilarity in the field. Remember that it is a very severe strain upon the patience and spirits of any one, to be urged to rapid labour of precisely the same description day by day, week by week, month by month. Let there be refreshments at the baskets, a dish of hot coffee in a cool morning, or a pail of buttermilk in a hot afternoon, or a tub of sweetened water, or a basket of apples."

As a rule the cotton gathered on one farm, which has, generally speaking, had something like uniformity in method of cultivation, will produce cotton varying very little in quality and weight.

Hence on large farms there will be something like uniform quality of cotton produced. It will, however, be clear to the general reader that on the small farms of India, say where sufficient cannot be gathered on one farm, or perhaps on a few farms, to make one bale, there will not be that uniformity which is desirable, hence Indian cotton, especially of the poorer types, varies a great deal more than the American varieties. When the hands have gathered sufficient to fill the carts drawn in America usually by mules, and in India by oxen, the cotton is taken to houses in which the seeds are separated from the fibre. This process is called "ginning."

It is astonishing to find how tenaciously the fibres cling to the seed when an attempt is made to separate them. At first much loss was occasioned because of the brutal methods employed, and now even with very much more perfect machinery a good deal of the cotton fibre is injured in the ginning process.

FIG. 8.—Indian women with roller gin.

At present, most of the cotton produced in various parts of the world is ginned by machinery, though in India and China foot gins and other primitive types are still employed.

It should be stated that where a large production of cotton is desired the foot gin or even what is known as the "Churka gin" (which consists of a couple of rollers turned by hand) is never employed. Only a few pounds a day of cotton can be separated from the seeds when this method is adopted.

The following extract from a lecture by the late Sir Benjamin Dobson will be of interest here, as showing what is done at an American ginnery:

"The farmer brings the cotton to the mill in a waggon, with mules or oxen attached; the cotton is weighed, and then thrown out of the waggon into a hopper alongside. From this hopper it is taken by an elevator, or lift, either pneumatic or mechanical, and raised to the third story of the ginning factory. There it is delivered into another part of the room until required. When the cotton is to be ginned it is brought by rakes along the floor to an open sort of hopper or trunk, and from here conveyed to the gins below by travelling lattices.

"In the factory of which I am speaking there were six gins, all of them saw-gins. Each gin was provided with a hopper of its own, and the attendant, when any hopper was full, could either divert the feed to some other gin, as he required, or stop it altogether. The gins produced from 300 pounds to 350 pounds per hour. The cotton is dropped from the condenser, in front of the gin, upon the floor close to the baling press, into which it is raked by the attendant and baled loosely, but only temporarily. The seed falls into a travelling lattice, and is conducted to a straight cylindrical tube, in which works a screw. This takes it some one hundred yards to the oil mill. There the seed is dropped into what are known as 'linting' machines, and as much as possible of the lint or fibre left upon the seed is removed.

"These linting machines—practically another sort of gin—deliver the cotton or waste in a kind of roll, which is straightway put behind a carding engine. Coming out of the carding engine it is made into wadding by pasting it on cardboard paper, for filling in quilts, petticoats, and for other purposes. When the seed has passed the linting machine, it is taken, still by a lattice, to a hulling machine. This machine will take off the outside shell, which is passed to one side, while the green kernel of the seed goes down a shoot. The seed fills certain receptacles placed in the oil press, and is submitted to a hydraulic press. The result is a clear and sweet oil, which I am credibly informed is sold in England and other countries under the name of 'olive oil.' The remains of the green kernel are then pressed into what are termed cattle cakes, or oil cakes, for feeding cattle."

But the reader is probably asking, what is a gin like?

The illustration seen in Fig. 9 is a gin which goes by the name of the "single-acting Macarthy gin," so called because it has only one oscillating blade for removing the fibre from the seed. The back of the machine is shown in the figure. This process at the best is a brutal one, especially when certain gins are employed, but the one figured here is considered to do little damage to the fibre when extracting the seed.

The gin shown in Fig. 9 is of simple construction, consisting of a large leather roller about 40 inches in length and 5 in diameter. "The roller is built up by means of solid washers, or in strips fastened on to wood, against which is pressed a doctor knife.

"The cotton is thrown into a hopper, and, falling, is seized by the friction of the leather and drawn between the doctor knife and the leather surface. Whilst this is taking place, there is a beater knife which is reciprocated at a considerable speed and which strikes the seed attached to the cotton drawn away by the leather roller. The detached seed will then fall through a grid provided for the purpose. A single-action gin should produce about 30 pounds of cleaned cotton per hour."

FIG. 9.—Single-acting Macarthy gin.

Another gin which does considerable damage to fibre, especially if it be over-fed, is still in use in the States. This was the invention of an American named Eli Whitney, and has been named a "saw-gin."

If the reader can imagine a number of circular saws (such as are to be seen in a wood-sawing mill) placed nearly together on a shaft to form an almost continuous roller, he will have a good idea of what the chief part of a saw-gin is like.

As the cotton is fed to the machine, the saws seize it and strip the cotton from the seeds, which fall through grids placed below the saws. The cotton is afterward stripped from the saws themselves by means of a quickly revolving brush which turns in the opposite direction to the saws. This gin is best suited to short stapled cottons, especially such as are grown in the States. For the longer fibred cotton this gin is not well adapted, much injury resulting to the cotton treated by it.

After the cotton is ginned, it is gathered into bundles and roughly baled. When a sufficient quantity has been so treated, it is carried to the "compressors," where the cotton undergoes great reduction in bulk as a result of the enormous pressure to which it is subjected.

For the general reader it will scarcely be necessary or wise to describe a "cotton press" in detail. Let it suffice to say that by means of a series of levers—in the Morse Press seven are used—tremendous pressure can be obtained. Thus for every 1 pound pressure of steam generated there will be seven times that pressure, if seven levers are used. When 200 pounds pressure of steam is up, there will be 1400 pounds pressure per inch on the cotton. So great is the pressure exerted that a bundle of cotton coming to the press from the ginnery, 4 feet in depth, is reduced to 7 inches when drawn from the compressor. While in the press iron bands are put round the cotton, and readers will have frequently seen cotton on its way to the mills having these iron bands round it.

The following table shows the number of bands which are found on bales coming to England from cotton-growing countries:—

	No. of bands.	Weight in lbs.
American bale	6 or 7	500
Egyptian "	11	700
Indian "	13	390
Turkish "	4	250-325
American Cylindrical bale	—	420-430
Brazilian	—	175-220

Within the last few years an entirely new industry has been started in some of the Southern States of America.

Up to recently the bales sent to European countries from America were all of the same type as shown by the centre bale in Fig. 10.

FIG. 10.—**Bales from various cotton-growing countries.**

Now a vast quantity of cotton is being baled in the form as shown in Fig. 11, and what are known as cylindrical bales are being exported in large numbers. In the "Round Bale" Circular of the American Cotton Company, it is stated that from the 21st November, 1896, to January 2nd, 1897, no less than 1443 round bales were turned out of the factory at Waco in Texas. The total weight of these bales was 614,832 pounds, giving an average of 426 pounds per bale.

By means of a press the cotton is rolled into the form as shown in the illustration. The press makes a bale 4 feet long and 2 feet in diameter and weighs over 35 pounds per cubic foot or 50 per cent. denser than the bale made under the system as shown in Fig. 10.

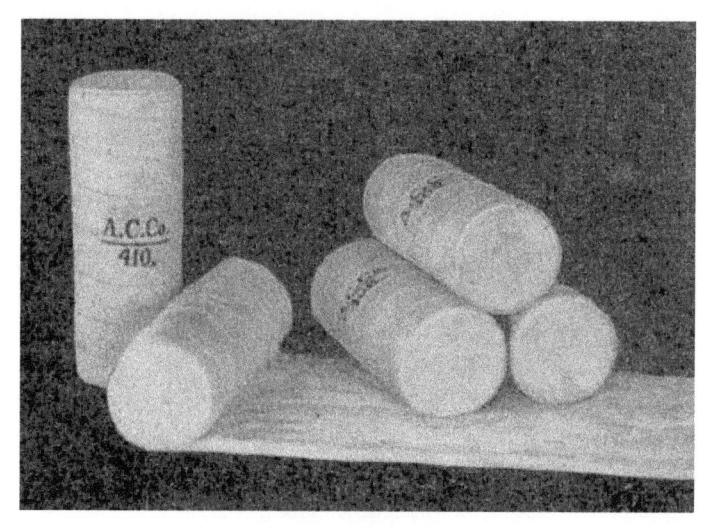

FIG. 11.—Cylindrical rolls of cotton.

It is claimed for this new system that the regularity of the size of the bale, 4 x 2 feet, makes it pack much closer than the irregular turtle-backed bales as usually made on the old system.

Under the new style the cotton is pressed gradually and not all at once. For this reason it is claimed that the fibre is not injured and the cotton arrives at the mill with the fibre in as good condition as when it left the gins.

"Bagging and ties are entirely dispensed with, as the air is pressed out of the cotton and it has no tendency to expand again, and the covering needed is only sufficient to keep the cotton clean."

From a number of experiments it is proved that the "round bale" is both fireproof and water proof.

From the illustration of the round bale shown in Fig. 11, it will be seen how readily this new form of bale lends itself to greatly aiding the operatives in the opening processes in the mill. The roll which lies on the floor like a roll of carpet could be so fixed that the cotton could be fed to the opener by being unrolled as shown in the illustration.

At present the round bale system is not popular and it remains to be seen whether it will commend itself to cotton spinners.

CHAPTER VI.

MANIPULATION OF COTTON IN OPENING, SCUTCHING, CARDING, DRAWING, AND FLY-FRAME MACHINES.

Before attempting to give the readers of this story an insight into the various operations through which cotton is made to pass, it may be advisable to briefly enumerate them first.

On the field there are the operations of collecting and ginning, that is, separating the raw cotton from the seeds. To the stranger it is very astonishing that as many as 66 to 75 pounds of seed are got from every 100 pounds of seed cotton gathered. Then in or near the cotton field the process of baling is carried out. Thus there are collecting, ginning and baling, as preliminary processes.

When the cotton arrives in bales at the mill (see Fig. 10), in which it is to be cleaned, opened and spun, it is first weighed and a record kept.

In the mill the first real operation is the taking of quantities of cotton from different bales of cotton from various countries, or different grades from the same country, and "mixing" so as to secure a greater uniformity in the quality of the yarn produced. In this process it is now the common practice to use a machine termed the "Bale Breaker," or "Cotton Puller."

The second important process carried out in the mill is "opening." By this the matted masses of cotton fibres are to a great extent opened out, and a large percentage of the heavy impurities, such as sand, shell, and leaf, fall out by their own weight. It is now also usual at this stage to form the cotton into a large roll or sheet called the "lap."

Immediately following the "opening" comes "scutching," which is merely a continuation of the work performed by the "opener," but done in such a way that greater attention is bestowed upon the production of an even sheet or "lap" of cotton.

The cotton at this stage is practically in the same condition as it was when first gathered from the tree in the plantation.

Carding comes next in order, and it should be observed that this is one of the most beautiful and instructive operations carried on in the mill.

The process of opening out the cotton is continued in this operation to such an extent that the fibres are practically *individually separated*, and while in this condition very fine impurities are removed, and many of the short and unripe fibres which are always more or less present are removed. Before leaving the machine the fibres are gathered together again in a most wonderful manner and converted into a "sliver," which for all the world looks like a rope of cotton, a little less than an inch in diameter.

In most mills "drawing" succeeds "carding," this operation having for its object (1) the doubling together of four to eight slivers from the card and attenuating them to the dimension of one so as to secure greater uniformity in diameter. (2) The reduction of the crossed and entangled fibres from the card into parallel or side by side order.

After "drawing," the cotton is brought to and sent through a series of machines termed "Bobbin and Fly Frames." There are usually three of these machines for the cotton to pass through, to which are given the names of "Slubbing," "Intermediate," and "Roving" Frames.

Their duties are to carry on the operation of making the sliver of cotton finer or thinner until it is ready for the final process of spinning, and incidentally to add to the uniformity and cleanliness of the thread of cotton.

The final process of spinning is chiefly performed on one of two machines, the "Mule" and the "Ring Frame," either of which makes a thread largely used without further treatment in a spinning mill.

Sometimes, however, the thread is further treated by such operations as doubling, reeling, gassing, etc. It should be added that in the production of the finest and best yarns an important process is gone through, named "combing."

This may be defined as a continuation of the carding process already named before to a much more perfect degree. The chief object is to extract all fibres below a certain required length, and reject them as waste. There is as much of this latter made at this stage of manufacture as that made by all the other machines put together, that is, about 17 per cent. Of course it will be readily seen that this is a costly operation and is limited entirely to the production of the very best and finest yarns.

This process necessitates the employment of a machine called a "Sliver Lap" and sometimes a "Ribbon Lap Machine" in order to put the slivers from the carding engine into a small lap suitable for the "creel" of the "Combing Machine."

Cotton Mixing and the Bale Breaker.—As before stated, the first operation in the mill is the opening out of bales of raw material and making a "mixing." Of course the weight of the bale is ascertained before it is opened.

All varieties of cotton vary in their commercial properties, this variation being due to a number of causes. From a commercial value point of view, there is an enormous difference between the very best and the very worst cottons; so much so, indeed, that they are never blended together. Between these two extremes there is a well-graded number of varieties and classifications of cotton, and some approximate so closely to others in quality, that they are often blended together in the "mixing."

Further than this, the same class of cotton often varies in spinning qualities from a number of circumstances that need not here be named. This is, however, an additional reason why cotton from various bales should be blended together in order to secure uniformity.

A cotton "mixing" may be described as a kind of "stack," resembling somewhat the haystack of the farm yards.

The method usually pursued in making this mixing is somewhat as follows:—A portion of cotton from a certain bale is taken off and spread over a given area of floor space. Then a similar portion from another bale is placed over the first layer already lying on the floor.

The same operation is followed with a third and fourth layer from different bales, and so on with as many bales as the management consider there are variations in quality, the larger the mixing the better for securing uniformity of yarn.

When it is desired to use the cotton, it should be pulled down vertically from the face of the "mixing," so as to secure a fair portion from each bale composing the mixture. Before spreading the cotton out it is usually pulled into pieces of moderate size by the hands of the operative.

During recent years it has become the very general practice to use a small machine called the "Bale Breaker" or "Cotton Puller," and to have also working in conjunction with this machine long travelling "lattices" called "mixing lattices." These perform the operation of "pulling" and "mixing" the cotton much more quickly and effectively than by hand labour.

The "Cotton Puller" or "Bale Breaker" (see Fig. 12) simply consists, in its most useful form, of four pairs of coarsely fluted or spiked rollers of about 6 inches diameter with a feed apron or lattice such as is shown in the illustration.

FIG. 12.—Bale breaker or puller.

The method adopted with the "Bale Breaker" and "mixing lattices" in use is as follows:—

The various bales of cotton intended for "mixing" are placed very near to the feed apron of the Bale Breaker, and a layer from each bale in succession is placed on the apron. The latter feeds the cotton at a slow rate to the revolving rollers of the machine, and as each pair of top and bottom rollers that the cotton meets is revolving more rapidly than the preceding pair, the result is a pulling asunder of the cotton by the rollers, into much smaller pieces, quite suitable for the next machine. The Bale Breaker delivers the cotton upon long travelling aprons of lattice work, which carry the cotton away and deposit it upon any desired portion of the floor to form the "mixing."

Opening.—The name of the next process, viz., "opening," has been given it because its primary function is "to open" out the cotton to such an extent that the greater bulk of the seed, leaf, sand, and dust is readily extracted. The details of this machine and indeed practically of all machines used in cotton spinning, vary so much with different makers, that it would be utterly out of place to deal with them here, so that it may be said at once, that all such points are entirely omitted from this treatment of the subject.

The essential and principal portions of the machines are practically identical for all makers, and it is with these only that it is proposed to deal, taking in all cases the best present-day practice.

The opener, then, is a very powerful machine, being in fact the most powerful used in cotton spinning, and the most important feature of the machine is the employment of a strong beater, to which is fitted a large number of iron or steel knives or spikes. These beat down the cotton and open it at a terrific rate, the beater having a surface speed of perhaps 4000 feet a minute. Various fans, rollers, and other parts are employed to feed the cotton to the beater, and to take it away again after treatment. It will perhaps best serve the purpose of our readers if the passage of the cotton be described through an opener of the most modern and approved construction, dealing with the subject in non-technical terms.

With this object in view, take for example what is termed "The Double Cotton Opener" with "Hopper Feed Attachment." This machine is shown in Fig. 13.

FIG. 13.—"Double opener" with "hopper feed."

The Hopper Feed is about the most recent improvement of any magnitude generally adopted in cotton spinning mills. It is an attachment to the initial or feed end of an opener with the object of feeding the cotton more cheaply and effectively than it can be done by hand.

It may be said to consist of a large iron feed box, into which the cotton is passed in considerable quantities at one time. At the bottom of the feed box, or hopper, is a travelling apron which carries the cotton forward, so as to be brought within the action of steel pins in an inclined travelling apron or lattice. This latter carries the cotton upwards, and special mechanism is provided in the shape of what is termed an "Evener roller," to prevent too much cotton going forward at once.

The cotton that passes over the top of the inclined lattice or apron is stripped off by what is denominated the stripping roller, and is then deposited on the feed apron of the opener, where formerly it was placed by hand.

It may be said that one man can feed two machines with Hopper Feeds as against one without them, and in the best makes the work is done more effectively.

The feed lattice of the opener carries the cotton along to the feed rollers, which project it forward into the path of the large beater. It is here that the opening and cleaning actions are chiefly performed.

The strong knives or spikes of the beater break the cotton into very small portions indeed, and dash it against "cleaning bars" or "grate bars" specially arranged and constructed. Through the interstices of these bars much of the now loosened seed and dirt present in the cotton passes into a suitable receptacle, which is afterward cleaned out at regular intervals.

The opened and cleaned cotton is taken away from the action of the beater by an air current produced by a powerful fan. This latter creates a partial vacuum in the beater chamber by blowing the air out of certain air exit trunks specially provided. To supply this partial vacuum afresh, air can only be obtained from the beater chamber, and the air current thus induced, takes the cotton along with it, and deposits it in the form of a sheet upon what are termed "cages" or "sieve cylinders."

These are hollow cylinders of iron or zinc perforated with a very large number of small holes through which the air rushes, leaving the cotton, as it were, plastered on the outer surfaces of the cages.

It is usual to have a pair of these cages, working one over the other like the pair of rollers in a wringing machine.

The cotton now passes between two pairs of small guide rollers, and is fed by the second pair to a second beater, but of very different construction from the first one.

This consists of two or three iron or steel blades extending the full width of the machine and carried by specially constructed arms from a strong central shaft.

The edges of these beater blades are made somewhat sharp, and they strike down the cotton from the feed roller at the rate of 2000 or more blows per minute.

This of course carries the opening work of the cotton of the first beater to a still further degree, and as in this case the cotton is also struck down upon "beater bars" or cleaning bars, a further quantity of loosened impurities passes through the bars. As before, another powerful fan creates an air current by which the cotton is carried away from the beater and placed upon a pair of "Cages." From this point the cotton is conducted in the form of a sheet between four heavy calender or compression rollers, the rollers being superimposed over each other, and the cotton receiving three compressions in its passage.

This makes a much more solid and tractable sheet of cotton, and it is now simply wound upon an iron roller in the form of a roll of cotton termed a "lap," being now ready for the subsequent process, as shown in the illustration (Fig. 14).

FIG. 14.—Scutching machine with "lap" at the back.

Scutching.—This term obviously means beating, and the process itself is simply a repetition of the opening and cleaning properties of the opener, these objects being attained to a greater degree of perfection. For the best classes of cotton it is often deemed sufficient to pass it through the opener alone, and then to immediately transfer the lap to the process of carding. For some cottons it is the practice to pass the cotton through two scutchers in addition to the opener, while in other cases it is the practice to use one scutcher only in addition to the opener.

In the scutcher it is the most common practice to take four laps from the opener and to place them in a specially constructed creel and resting on a travelling "lattice" or apron. By this they are slowly unwound and the four sheets are laid one upon another and passed in one combined sheet, through feed rollers, to a two or three bladed beater, exactly like the second one described when treating upon the double opener. Also, exactly in the same manner, a lap is formed ready for the immediately succeeding process of carding. In the scutcher the doubling of four laps together tends to produce a sheet of cotton more uniform in thickness and weight than that from the opener. This object of equality of lap is also invariably aided by what are termed Automatic Feed Regulators, which regulate the weight of cotton given to the beater to something like a continuous uniformity. The action is clearly seen in the illustration.

Carding.—By many persons this is deemed to be the most important operation in cotton spinning. Its several duties may be stated as follows:—

1. The removal of a large proportion of any impurities, such as broken leaf, seed and shell, that may have escaped the previous processes. It may usually be deemed to be the final process of cleansing.

2. To open out and disentangle the clusters of fibres into even greater individualisation than existed when first picked, and to leave them in such condition that the subsequent operations can easily draw them out, and reduce them to parallel order.

3. The extraction of a good proportion of the short, broken and unripe fibres, present more or less in all cottons grown, and practically worthless from a manufacturing point of view.

4. The reduction of the heavy sheet or lap of cotton from the scutcher, into a comparatively light and thin sliver. Ordinarily, one yard of the lap put up behind the card weighs more than 100 times as heavy as the sliver delivered at the front of the card.

There are several varieties of Carding Engine, but in each case nearly all the essential features are practically the same in one card as in another. At the present time, the type of Carding Engine which has practically superseded all others is denominated the "Revolving Flat Card." This Card originated with Mr. Evan Leigh, of Manchester, and after being in close competition with several other types has almost driven them out of the market. Of course it has been considerably improved by later inventors, and various machine makers have their own technical peculiarities.

In the illustration seen in Fig. 15 there is conveyed an excellent idea of the appearance of the heavy lap of cotton as it is placed behind the Carding Engine, and of the manner in which the same cotton appears as a "sliver" or soft strand of cotton as it issues from the front of the same machine, and enters the cylindrical can into which it is passed, and coiled into compact layers, suitable for withdrawal at the immediately succeeding process.

FIG. 15.—Two views of the carding engine: upper view, cotton entering; lower view, cotton leaving.

In the main, the parts which operate upon the cotton fibres in their passage through this machine consist of a number of cylinders or rollers of various diameters, but practically equal in width. Some of these rollers are merely to guide and conduct the cotton forward, but the more important are literally bristling all over with a vast number of closely set and finely drawn steel wire teeth, whose duty it is to open, and comb out, and clean the fibers as they pass along.

To begin with, the "lap" or roll of cotton is placed behind the machine so as to rest on a roller of 6 inches in diameter, which slowly unwinds the lap at the rate of about 9 inches per minute, by frictional contact therewith.

Here, it may be said that the width of this and other chief rollers and cylindrical parts of the card may be about 38 inches or 40 inches wide, there being a tendency to make present-day Carding Engines rather narrower than formerly, in order to give greater strength to certain parts.

From the lap roller the sheet of cotton is conducted for about 8 inches over a smooth feed plate, and then it goes underneath a fluted roller of 2¼ inches diameter, termed the feed roller, having practically the same surface speed as the lap roller, or possibly a small fraction more to keep the cotton lap tight.

At this stage the actual work of the Carding Engine may be said to commence. While the feed roller and the feed plate hold the end of the sheet of cotton and project it forward at the slow rate of 8 or 9 inches per minute, this projecting end of the lap becomes subject to the action of a powerful roller or beater termed the taker-in or licker-in.

The most recent and improved construction of this roller is termed the Metallic Taker-in, and it is covered all over with strong steel teeth shaped something like those of a saw. It is about 9 inches in diameter, and its strong teeth strike the cotton down from the feed roller with a surface speed of nearly 1000 feet per minute.

It is at this stage that the bulk of the heavier impurities still found in the cotton are removed, as these fall through certain grids below the taker-in immediately they are loosened from the retaining fibres by the powerful teeth of the taker-in.

The great bulk of the cotton fibres, however, are retained by the teeth of the taker-in and carried round the under side to a point where they are exposed to the action of the central and most important part of every Carding Engine, viz., the main "cylinder." The licker-in contains about twenty-eight teeth per square inch, but the "cylinder" is the first of the parts that the cotton arrives at, previously referred to as being covered with a vast number of closely set steel wire teeth.

Just to convey an idea of this point to the uninitiated reader, it may be said that it is quite common to have on the "cylinder" as many as 600 steel wire teeth in one square inch. For a cylinder 40 inches wide and 50 inches diameter, this works out to the vast number of over 3,800,000 steel wire teeth on one cylinder, each tooth being about ¼ inch long, and secured in a cloth or rubber foundation before the latter is wound round the cylinder.

The steel teeth of the cylinder strip the fibres from the taker-in and carry them in an upward direction, the surface speed of the cylinder being over 2000 feet per minute.

Placed over the cylinder, and extending for nearly one-half of its circumference, are what are technically known as the "flats."

These are narrow iron bars, each about 1⅜ inches wide; each being covered with steel wire teeth in the same manner as the cylinder; and each extending right across the width of the cylinder, and resting on a suitable bearing termed the "bend."

They are formed into an endless chain containing about 108 "flats," but only about 44 of which are in actual work at one time; this endless chain of flats being given a slow movement of about 3 inches per minute.

Here it may be said that the various working parts are set as close as possible to each other without being in actual contact, the usual distance being about 1/143rd of an inch determined by a specially constructed gauge, in the hands of a skilled workman.

The steel teeth of the flats, being set very close to those of the cylinder, catch hold of and retain a portion of the short warty fibres and fine impurities that may be on the points of the cylinder teeth, the amount of this reaching about 3 per cent. of the cotton passed through the machine. In addition to this the teeth of the flats work against those of the cylinder so as to exercise a combing action on the cotton fibres.

Having passed the "flats," the cotton is deposited by the cylinder on what is termed the doffer. This is a cylindrical body, exactly similar to the main "cylinder" excepting that it is only about half the diameter, say 24 inches. Its steel wire teeth are set in the opposite way to those of the cylinder, and its surface speed is only about 75 feet per minute. These two circumstances acting together enable it to take the cotton fibres from the main cylinder.

The operations of carding may now be said to be practically performed, as the remaining operations have for their object the stripping, collecting, and guiding of the cotton into a form suitable for the next succeeding processes. The fleece of cotton is stripped from the doffer by the "Doffer Comb," which is a thin bar of steel, having a serrated under edge, and making about 1600 beats or strokes per minute. From this point cotton is collected into the form of a loose rope or "sliver," and passed first through a trumpet-shaped mouth, and then through a pair of calender rollers about six inches wide and four inches in diameter.

FIG. 16.—Lap, web, and sliver of cotton.

Finally, the sliver of cotton is carried upward, as shown in the illustration (Fig. 15), and passed through special apparatus and deposited into the can, also shown. This latter is about 10 inches in diameter and 36 inches in length, and the whole arrangement for depositing the cotton suitably into the can is denominated the "Coiler." In the next illustration (Fig. 16) are shown three forms in which the cotton is found before and after working by the Carding Engine. That to the left is the lap as it enters, the middle figure is part of the web as it comes from the doffer, and that to the right is part of a coil of cotton from the can.

Such is a brief description of the most important of the preparatory processes of cotton spinning. There are innumerable details involving technical knowledge which fall outside the province of this story.

Drawing Frames.—It is a very common thing for a new beginner in the study of cotton spinning to ask—what is the use of the drawing frame? As a matter of fact, the unpractised eye cannot see any difference between the sliver or soft rope of cotton as it reaches, the drawing frame and as it leaves the frame.

The experienced eye of the practical man can, however, detect a wonderful difference.

It has been shown that the immediately preceding operation of carding—amongst other things—reduces the heavy lap into a comparatively thin light sliver; thus advancing with one great stride a long way toward the production of the long fine thread of yarn ready for the market.

No such difference can be perceived in the sliver at the drawing frame. This machine is practically devoted to improving the thread finally made in two distinct and important ways.

1. The fibres of cotton in the sliver, as they leave the Carding Engine, are in a very crossed and entangled condition, not at all suited to the production of a strong yarn by the usual processes of cotton spinning. The first duty of the drawing frame may be said, therefore, to be the laying of the fibres in parallel order to one another, by the action of the drawing rollers.

2. The sliver of cotton, as it leaves the card, is by no means sufficiently uniform in weight per yard for the production of a uniform and strong finished thread. It will easily be conceived by the readers of this story of the cotton plant that the strength of any thread is only that of its weakest portions.

Take a rope intended to hold a heavy weight suspended at its lower end, and assume it to be made of the best material and stoutest substance, but to contain one very weak place in it; this rope would practically be useless, because the strength of the rope would only be that of the weakest part.

The drawing machine in cotton spinning aims at removing the weak places in cotton thread, thus making the real strength of the thread vastly greater than it would otherwise be.

The method by which these important objects are attained may be briefly explained as follows:—

From four to eight, but most usually six, cans of sliver from the previous machine are placed behind the frame, and the ends of the slivers conducted over special mechanism within the range of action of four pairs of drawing rollers. This passage of the cotton is shown very clearly in Fig. 17.

The top rollers are made of cast iron, covered with soft and highly finished leather made from sheepskins, the object of this being to cause the rollers to have a firm grip of the cotton fibres, without at the same time injuring them. The bottom rollers are of iron or steel, made with longitudinal flutes or grooves, in order to bite the cotton fibres firmly on the leathers of the top rollers. In order to assist the rollers in maintaining a firm grip of the fibres the top rollers are held down by somewhat heavy weights.

The action of the drawing rollers will be adequately discussed later in this story, when dealing with the inventions of Lewis Paul and Sir Richard Arkwright, and need not be enlarged upon at this stage.

It will be sufficient, therefore, to say that, assuming that six slivers are put up together at the back of the frame, the "draft" or amount of drawing-out between the first and second pairs of rollers the cotton comes to, may be about 1.3, between the second and third pairs 1.8, and between the third and fourth pairs 2.6. These three multiplied together give a total draft of slightly over 6.

In other words, assuming that 1 inch of cotton be passed through the first pair of rollers, the second pair will immediately draw it out into 1.3 inches; the third pair will draw out the same portion of cotton into 1.3×1.8 inches = 2.34 inches, and the fourth or last pair of rollers will draw out the same portion of cotton into 2.34×2.6 inches = 6.084.

FIG. 17.—Drawing frame showing eight slivers entering and one leaving the machine.

The six slivers put up at the back are therefore drawn out or attenuated to the dimensions of one by the rollers, and then at the delivery side of the machine the six slivers are united into one sliver, and arranged in beautiful order inside a can exactly as described for the Carding Engine.

Now it is in the doubling together and again drawing-out of the slivers of cotton that the two objects of making the fibres parallel and the slivers uniform are effected.

In the first place, even the uninitiated readers of this story may conceive that the combining of six slivers will naturally cause any extra thick or thin places in any of the individual slivers to become much reduced in extent by falling along with correct diameters of the other five slivers; and experience proves that such is the actual fact. In this way the slivers, or soft untwisted ropes of cotton, are made uniform.

It is perhaps not so easy to see how it is that drawing rollers make the fibres of cotton parallel. As a matter of fact, it may be said that as each pair of rollers projects the fibres forward, the next pair of rollers takes hold of the fibres and draws their front extremities forward more rapidly than the other pair will let the back extremities of the same fibres pass forward. It is this action often repeated that draws the fibres straight, or in other words, reduces them to a condition in which they are parallel to each other.

It is the usual practice to pass each portion of cotton through three separate frames in this manner, in immediate and rapid succession. The "slivers" or ropes of cotton made at the front of the first drawing frame, would be placed in their cans behind a second drawing frame and the exact process just described would be repeated. The same identical process would usually be performed yet a third time in order to secure the required objects with what is considered a sufficient degree of perfection.

After this the cotton is usually deemed to be quite ready for the immediately succeeding process of "slubbing."

Bobbin and Fly Frames.—The series of machines now to be dealt with, are distinguished more for their complicated mechanism in putting twist into the attenuated cotton and in winding it upon bobbins in suitable form for the immediately succeeding process, than for the action of the parts upon the cotton so as to render it better fitted for the production of strong, fine yarn.

The manner in which these machines perform a part in the actual production of a thread or yarn is practically a repetition of the work of the drawing frame, with the great difference that the strand or thin rope of cotton leaves each machine of the series in a thinner and longer condition than when it arrived.

This attenuation of the cotton roving is indeed the chief desideratum that bobbin and fly frames aim at, although they assist in making the strand of cotton more uniform by carrying still further to a limited extent the doubling principle so extensively utilised at the drawing frames.

The basis of the operations are again the drawing rollers, brought to such a state of perfection by Richard Arkwright, and here it may be useful to remind the readers of this story how superior in this respect of general adaption Arkwright's method of spinning was to that of Hargreaves'. It will be remembered that the latter named inventor utilised a travelling carriage, for drawing the cotton finer, while the former performed the same work by drawing rollers.

Although the travelling carriage principle was at one time somewhat largely utilised in preparing the rovings for the final process of spinning, it has long since entirely given way before the superior merits and adaptability of the drawing roller principle; and it is now this latter method which is universally employed.

It usually takes three bobbin and fly frames to make up what may be called a "set," each portion of the cotton roving passing through the three machines in succession. For low classes of yarn only two of these machines may be used, while for the finest yarns there are sometimes four used to make up the "set."

Of course, all the readers of this story must understand that in an ordinary-sized cotton spinning mill there will be many sets of these machines, just as there will be a large number of "carding engines" and "drawing frames," and mules. Bale brakers, openers and scutchers are so very productive that only a limited number is required as compared with the other machines already named.

Those of our readers who have studied the details of Arkwright's spinning frame, described in another chapter in this book, and have understood those details, will have a clear comprehension of the action of the parts and leading mechanical principles concerned in the operations of a modern bobbin and fly frame. Certainly there are some of the most difficult problems of cotton spinning involved in the mechanism of these machines, but these points are so highly technical that it is not intended to introduce them here.

The "set" of machines just named are usually known by the names "Slubber," "Intermediate or Second Slubber," and "Roving" Frames.

Nearly all the operations and mechanisms involved in one are almost identical in the others, so that a description of one only in the set is necessary, merely explaining that the parts of each machine the cotton comes to in the latter two of the set are smaller and more finely set than the corresponding parts of the immediately preceding machine.

Taking the Intermediate frame as a basis, the operation may be described as follows:—The bobbins formed at the slubbing frame are put in the creel of the Intermediate, as shown in the photograph (Fig. 18), each bobbin resting on a wooden skewer or peg which will easily rotate.

In order to increase the uniformity of the roving or strand of cotton, the ends from two of the slubbing rovings are conducted together through the rollers of the machine.

There are three pairs of these rollers, acting on the cotton in every way just as described for the drawing frame.

Although two rovings are put together behind the rollers, yet the "draft" or drawing-out power of the rollers is such, that the roving that issues from the front of the rollers is about three times as thin as each individual roving put up behind the rollers. This drawing-out action of the rollers need not be further dilated upon at this stage.

The points which demand some little attention at our hands, are the methods and mechanism involved in twisting the attenuated roving, and winding it upon bobbins or spools in suitable form for the next process.

FIG. 18.—Intermediate frame (bobbin and fly frame).

As regards twisting of the roving it must be distinctly understood that when the attenuated strand of cotton issues from the rollers of the first bobbin and fly frame, it has become so thin and weak that it can no longer withstand the requisite handling without being seriously damaged. Hence the introduction of "Twist," which is by far the most important strength-producing factor or principle entering into the composition of cotton roving and yarn.

Without twist there would be no cotton factories, no cotton goods; none of the splendid and gigantic buildings of one description or another which are found so plentifully intermingled with the dwellings and factories of large cotton manufacturing towns!

In a sense it is to this all-powerful factor of "twist" that all these buildings owe their existence, since it would be practically impossible to make a thread from cotton fibres without the assistance of "twist" to make the fibres adhere to each other. Hence there could be none of that wealth which has caused the erection of these buildings.

This is true in a double sense, since we have both the natural twist of the cotton fibres and the artificial twist introduced at the latter processes of cotton spinning, in order to make individual fibres and aggregations of fibres adhere to each other. What is termed the natural twist of the fibres may average in good cottons upwards of 180 twists per inch, while the twists per inch put into the finished threads of yarn from those fibres may vary, say, between 20 and 30 twists per inch.

In all the fly frames, therefore, this artificial twist is invariably and necessarily put into the roving. As the cotton leaves the front or delivery rollers, each strand descends to a bobbin of from 8 to 12 inches long, upon which it is wound by special mechanism. As in Arkwright's frame, this bobbin is placed loosely upon a vertical "spindle," and upon the latter is fitted a "flyer," whose duty it is to guide the cotton upon the bobbin.

The primary duty of the spindle is to insert the "twist" which has been shown to be so necessary to give sufficient strength to the roving.

Let any reader of this story hold a piece of soft stuff in one hand while with the other hand he rotates or twists the roving and he will have an idea of the method and effect of twisting (see Fig. 19).

Without going into minute details we may say that the practical effect is that, while the roving is held firmly by the rollers, it is twisted by means of its connection at the other end to the rotating bobbin, spindle and flyer. The twist runs right from the spindle along the 6 to 12 inches of cotton that may extend from the spindle top to the "nip" of the rollers, thus imparting the requisite strength to the roving as it issues from the rollers. The mechanism for revolving the spindles is by no means difficult to understand, simply consisting of a number of shafts and wheels revolved at a constant, definite and regulated speed per minute.

Not only is it necessary to provide special apparatus for twisting the cotton at the bobbin and fly frames, but also very complicated and highly ingenious mechanism for winding the attenuated cotton in suitable form upon the bobbins. Indeed it is with this very mechanism that some of the most difficult problems of cotton spinning machinery are associated.

Although the cotton at this stage is strengthened by twist, yet it is extremely inadvisable and practically inadmissible to insert more than from 1 to about 4 twists per inch at any of these machines, so that at the best the rovings are still very weak.

If too much twist were inserted at any stage, the drawing rollers of the immediately succeeding machine could not carry on the attenuating process satisfactorily.

This winding problem was so difficult that it absolutely baffled the ingenuity of Arkwright and his contemporaries and immediate successors, and it was not until about 1825 that the difficulties were solved by the invention of the differential winding motion by Mr. Holdsworth, a well-known Manchester spinner, whose successors are still eminent master cotton spinners.

This winding motion is still more extensively used than any other, although it may be said that quite recently several new motions have been more or less adopted, whose design is to displace Holdsworth's motion by performing the same work in a rather more satisfactory manner.

In these pages no attempt whatever will be made to give a technical explanation of the mechanism of the winding motion. It may be said that it was a special application of the Sun and Planet motion originally utilised by Watt in his Steam Engine, for obtaining a rotary motion of his fly-wheel.

Sufficient be it to say that this "Differential Motion," acting in conjunction with what are termed "Cone drums," imparts a varying motion to the bobbins upon which the cotton is wound, in such a manner that the rate of winding is kept practically constant throughout the formation of the bobbins of roving, although the diameters of the latter are constantly increasing.

The spindles and bobbins always rotate in the same direction, but while the revolutions per minute of the spindles are constant, so as to keep the twist uniform, those of the bobbins are always varying, in order to compensate for their increasing diameters or thicknesses of the bobbins. The delivery of cotton from the rollers is also constant and the mechanism required to operate them is exceedingly simple.

A vast number of details could easily be added respecting the operations performed by the bobbin and fly frames, but further treatment is deemed unnecessary in this story.

CHAPTER VII.

EARLY ATTEMPTS AT SPINNING, AND EARLY INVENTORS.

There can be no better illustration of the truth of the old saying, that "Necessity is the mother of invention," than to read the early history of the cotton manufacture, and the difficulties under which the pioneers of England's greatest industry laboured.

The middle years of the eighteenth century act as the watershed between the old and the new in cotton manufacture, for up to 1760 the same type of machinery was found in England which had existed in India for centuries. But a change was coming, and as a greater demand arose for cotton goods, it became absolutely necessary to discover some better way of manipulating cotton, in order to get off a greater production.

"When inventors fail in their projects, no one pities them; when they succeed, persecution, envy, and jealousy are their reward." So says Baines, and it would appear, from reference to the history of the cotton industry, to be only too true. Certain it is, that the early inventors of the machinery for improving cotton spinning did not reap the advantages which their labours and inventions entitled them to. They ploughed and sowed, but others reaped.

Among the most celebrated of the early inventors, the following stand out in great prominence—John Kay, Lewis Paul, John Wyatt, Richard Arkwright, Thomas Highs, James Hargreaves, and Samuel Crompton.

When and how spinning originated no one can say, though it can be traced back through many, many centuries. Several nations claim to have been the first to discover the art, but when asked for proof the initial stages are greatly obscured by impenetrable clouds of mystery.

For example, the Egyptians credit the goddess Isis with the discovery, the Greeks Minerva, the Chinese the Emperor Yao. It is related of Hercules, that, when in love with Omphale, he debased himself by taking the spindle and spinning a thread at her feet. This form of work was considered to belong only to women, and by spinning for her in this position he was thought to have greatly humiliated himself.

If Hercules were back again, and could stand between two modern mules and see the men and boys engaged in spinning hundreds of threads *at once*, no doubt he would wonder, just as we do to-day at his fabled feats.

It is not difficult to imagine that very early on in the world's history the twisting together of strands of wool and cotton would force itself upon the attention of the ancients. If the reader will take a little cotton wool in the left hand and by means of the first finger and thumb of the right take a few cotton fibres and gently twist them together and at the same time draw the thread formed outwards, it will be seen how very easy it is (from the nature of the cotton) to form a continuous thread.

What would very soon suggest itself would be something to which the thread, when twisted, could be fastened and, according to Mr. Marsden (who supposes the first spinner to have been a shepherd boy), a twig which was close at hand would be the very thing to which he could attach his twisted fibres. He also supposes that, having spun a short length, the twig by accident was allowed to dangle and immediately to untwist by spinning round in the reverse way, and ultimately fall to the ground.

He further adds, the boy would argue to himself "that if this revolving twig could take the twist out by a reversion of its movements, it could be made to put it in." This would be the first spinning spindle. The explanation is probably not very far wide of the mark.

A weighted twig or spindle would next be used, and as each length of spun thread was finished, it would be wound on to the spindle and fastened.

As it would be extremely awkward to work the fibre up without a proper supply, a bundle of this was fastened to the end of a stick and carried most probably under the left arm, leaving the right hand free, or in the belt, much in the same way as is done in some country districts in the North of Europe to-day.

The modern name for this stick is *Distaff*, a word which is derived from the Low German—*diesse*, the bunch of flax on a distaff, and *staff*. Originally it would be the staff on which the tow or flax was fastened, and from which the thread was drawn. The modern representative of the spindle with the twisted thread wound on it is the "*cop*," and the intermittent actions of first putting *twist* in the thread and then *winding* on the spindle, have their exact counterparts on the latest of the self-acting mules of to-day.

FIG. 19.—Twist put in cotton by the hand.

It may be interesting to note that St. Distaff's Day is January 7th, the day after the Epiphany, a church festival celebrated in commemoration of the visit of the Wise Men of the East to Bethlehem. As this marks the end of the Christmas festival, work with the distaff was commenced, hence the name, St. Distaff's Day.

It is also called "Rock Day," rock being another name for distaff. "Rocking Day" in Scotland was a feasting day when friends and neighbours met together in the early days of the New Year, to celebrate the end of the Christmastide festival.

The reign of Henry VII. is said to have witnessed the introduction into England of the spindle and distaff.

In process of time, the suspended spindle was superseded by one which was driven by mechanical means. Over and over again, the spindle, as it lay upon the floor, must have suggested that it could be made to work in that position, viz., horizontal. And so comes now a contrivance for holding the spindle in this position.

Mr. Baines, in his history of the cotton manufacture, gives a figure of an old Hindoo spinning wheel, and it is extremely likely that this very form of machine was the forerunner of the type which later on found its way into Europe. At the beginning of the sixteenth century what was known as the Jersey wheel came into common use. This machine is shown in Fig. 20.

Lying to the left hand of the woman in the illustration is a hand card. This consisted of square board with a handle, and was covered by fine wire driven in, so as to make what was really a wire brush. By means of this, the spinner was enabled to prepare her cotton, and she did with it (though not nearly so well) what is done by the Carding Engine of to-day, viz., fully opened out the fibres of cotton ready for spinning. Having taken the cotton from the hand cards, she produced at first a very thick thread which was called a *roving*. This she wound on a spindle, which was afterwards treated again on the wheel a second time, and drawn out still more, and then having the twist put in, it was made much thinner into so-called yarn. Only one thread could by this method be dealt with at a time by one person, but the main operations carried out on the old spinning wheel have their exact reproductions on the mule of to-day, viz.:—Drawing, Twisting and Winding.

FIG. 20.—**Jersey spinning wheel (after Baines).**

But still the process of evolution went on, and following quickly on the heels of the Jersey wheel is the Saxony or Leipsic wheel. Here for the first time is seen the combination of spindle, flyer and bobbin.

This machine was so arranged that by means of two grooved wheels of different diameters, but both driven by the large wheel similar to the one in the Jersey wheel, and which was operated by the spinner, two speeds were obtained. The bobbin was attached to the smaller, and the spindle, to which was fastened the flyer or "Twister," was driven by the larger of two wheels.

In this form of spinning machine, then, there were the following operations performed:—

By the spindle and flyer both revolving at the same velocity, the thread was attenuated and twisted as it was carried to the bobbin. This latter was, as already named, driven by the smaller of the two wheels and had a motion all its own, though much quicker than that of the spindle. In this way a bobbin of yarn was built up, and the Saxony wheel no doubt gave many fruitful ideas to the inventors who appeared later on, and who, by reason of their research and experiment, evolved the fly frames of to-day; this was notably so in the case of Arkwright.

There had been very great opposition to the introduction of cotton goods into England by manufacturers and others interested in the wool and fustian trade, and matters even got so bad that the British Parliament was foolish enough to actually pass an Act in 1720, prohibiting "the use or wear in Great Britain, in any garment or apparel whatsoever, of any printed, painted, stained, or dyed *calico*, under the penalty of forfeiting to the informer the sum of £5."

Just as though this was not sufficiently severe, it was also enacted that persons using printed or dyed calico "in or about any bed, chair, cushion, window-curtain, or any other sort of household stuff or furniture," would be fined £20, and a like amount was to be paid by those who sold the stuff.

There can be no doubt whatever, that this Act was designed to strike a death-blow at the cotton industry, which at this time was beginning to make itself felt in the commerce of the country. A curious exception should be mentioned here. Calico, which was all blue, was exempted from the provisions of this Act, as were also muslins, fustians and neck-ties. However, in 1736 this iniquitous piece of legislation was somewhat relaxed, and Parliament was good enough to decree in the year just named that it would be lawful for anyone to wear "any sort of stuff made of linen yarn and cotton wool manufactured and printed or painted with any colour or colours within the kingdom of Great Britain, provided that the warp thereof be entirely linen yarn."

Now as half a loaf is better than none, the cotton manufacturers received a direct impulse by the partial removal of the obnoxious restriction, and very soon the supply was far ahead of the demand.

Manufacturers were crying out constantly for more weight and better stuff, but how by the mechanical means at the disposal of the spinners were they to get it? Lancashire historians say that it was no uncommon thing for weavers to travel miles in search of weft, and then many of them returned to their looms with only a quarter of the amount they required.

Another cause which acted in the direction of increasing the demand for yarns and weft was the invention of the *flying shuttle* by John Kay about

1738. Previous to his time, the heavy shuttles containing the wefts were sent across the looms by two persons. Now, by his new shuttle he dispensed with the services of one of these artisans, and by means of his arrangement for quickly sending the shuttle along the lathe of the loom, much more cloth was produced. Poor Kay suffered much by the cruel persecution of his countrymen, who ignorantly supposed that in bringing his new shuttle to such perfection, they would be deprived *permanently* of their occupations, with nothing but starvation looking them in the face. Of course, nothing could be wider of the truth than this, but Kay had to flee his country, and died in poverty and obscurity in a foreign land. Still the shuttle continued to be used, for the makers of cloth had learned that increased production meant more work, and possibly greater profit, and though Kay disappeared, his works remained behind. The demand for weft grew more and more. It has been said that it is the occasion which makes the man, and not man the occasion. It was so in this case, for here was a cry for some mechanical means to be discovered for satisfying the ever-increasing demand for cotton weft. Hitherto single threads only had been dealt with on the spinning machines, but the same year witnessed the introduction of an invention which in a few years completely revolutionized the spinning industry, and which enabled one worker to spin hundreds of threads at once.

The year 1738, which witnessed the birth of Kay's invention, also saw that of Lewis Paul, an artisan of Birmingham. This was a new method of spinning by means of *Rollers*. It should be remembered that this was thirty years before Arkwright attempted to obtain letters patent for his system of spinning by rollers.

Most of the readers of this little book will have seen what is known in *domestic parlance* as a clothes-wringer. Here the wooden or rubber rollers, by means of weights or screws, are made to squeeze out most of the moisture which remains after the garment has left the washing-tub. Now if two sets of such rollers could be put together, so that in section the four centres would coincide with the four angular points of a square, and the back pair could be made to have a greater surface velocity than the front pair, this arrangement would give something like the idea which Paul had in his mind at that time. Why make the back pair revolve at a greater rate? For this reason, that as the cotton was supplied to the front pair, and passed on to the second, remembering that these are going at a greater rate, it follows that the cotton *would be drawn out* in passing from the first to the second pair. Had the rollers been both going at the same speeds, the cotton would pass out as it went in, unaffected. Now it was this idea which Paul practically set out in his machine. From some reason or other, Paul's right to this patent has been often called into question, and up to 1858 it was popularly supposed to have been the sole invention of John Wyatt of Birmingham. In the year named, Mr. Cole, in a paper read before the British Association, proved that Paul was the real patentee, and established the validity of his claim without doubt.

The two distinguishing features of Paul's Spinning Machine were: (1) by means of the rollers and flyers he performed the operations of drawing-out and twisting, which had hitherto been done by the fingers and thumbs of the spinners; and (2) he changed the position of the spindle itself from the horizontal to the vertical.

A glance at the Transactions of the Society for the Encouragement of Arts, Manufactures and Commerce, shows that this period (1760-1770) was most prolific of inventions specially relating to the various sections of the cotton industry. There were "improved spinning wheels," "a horizontal spinning wheel," and three other forms of "spinning machines" submitted to the above society between 1761 and 1767, in the hope of obtaining money grants in the shape of premiums, which had been offered to the best inventions for improving spinning machinery in general.

The above list does not however contain any reference to one improvement by James Hargreaves of Blackburn, Lancashire, to which in this story special mention must be made.

It appears that in 1764 or 1765 this individual had completed a machine for spinning eleven threads *simultaneously*; and five years later he had developed the machine to so perfect a state that he took out a patent for it, from which time it became known to the industrial world as a *Spinning Jenny*. His right to the patent has over and over again been challenged, and it has been alleged that Thomas Highs of Leigh, also in Lancashire, was the real inventor. Baines, in his "History of the Cotton Manufacture," is inclined to the view that Hargreaves was the first to perfect the machine known as the "Jenny" (see Fig. 21).

From whatever point of view Hargreaves' machine is looked at, it must be acknowledged to be a decided step forward in the direction of spinning machinery improvement.

The jenny was so unlike Arkwright's frame or Paul's, and preceded that of the former by some years, that its claim to originality can not be questioned. How the inventor came to produce his machine can not be stated, but it is reported that on one occasion he saw a single thread spinning wheel which had been accidentally knocked over, lying with the wheel and spindle free and both revolving.

If the reader will think for a minute it will be apparent that the horizontal position of the spindle would be changed to a vertical one, and Hargreaves argued if one spindle could revolve in that way, why should not eight or any number of spindles be made to work at the same time. How far he successfully worked out that idea will be seen if reference be made to the illustration of the jenny which is shown in Fig. 21.

After what has been said under the head of Carding, Drawing, and Roving, it will easily be understood when it is said that, unlike Arkwright's Machine, Hargreaves' Jenny could only deal with the cotton when in the state of *roving*, and it was the roving which this machine attenuated and twisted or spun into yarn.

If the reader will imagine he or she is standing in front of the jenny, the following description will be made much clearer:—

FIG. 21.—Hargreaves' spinning jenny (after Baines).

The rovings, which have previously been prepared, are each passed from the bobbins seen on the lower creel, through a number of grooves on one of the bars which run across the frame, as seen in the illustration. These rovings are next passed on to the spindles standing at the back of the frame and secured to them. A second bar in front of the one over which the rovings pass, acts as a brake and prevents, when in its proper position, any more roving being delivered, thus securing all between the spindles and the said bar. The wheel which is seen on the right of the jenny communicates with a cylinder by means of a strap or rope, and this cylinder in turning gives circular motion to the spindles which are connected with the cylinder by endless bands. On the spindle is the wharf, specially formed to allow the band to run without slipping.

The operations for a complete spinning of one delivery is described by Baines as follows:—

"A certain portion of roving being extended from the spindles to the wooden clasp, the clasp was closed, and was then drawn along the horizontal frame to a considerable distance from the spindles, by which the threads were lengthened out, and reduced to the proper tenuity; this was done with the spinner's left hand, and his right hand at the same time turned a wheel which caused the spindles to revolve rapidly, and thus the roving was spun into yarn. By returning the clasp to its first situation, and letting down a presser wire, the yarn was wound on the spindle."

Hatred and jealousy were immediately born when Hargreaves' splendid improvement became known, and, like poor Kay before him, he had to leave his native soil and get to some more secluded spot. He ultimately arrived in Nottingham, set at once to accommodate himself to his new environment, and soon entered into partnership with a Mr. James, and in 1770 took out a patent for his Jenny. In conjunction with his new partner, a mill was built, said to be one of the first, if not the first, spinning mill so called in this country.

Though it is stated by Arkwright that Hargreaves died in comparative obscurity and poverty, others say that this is not so; though he was not wealthy the evidence is sufficiently good to believe that he died in moderate circumstances.

The register of St. Mary's Parish, Nottingham, contains the following entry:—"1778, April 22, James Hargraves."

CHAPTER VIII.

FURTHER DEVELOPMENTS—ARKWRIGHT AND CROMPTON.

Whatever may be said in favour of other spinning machinery inventors, it is quite certain that when we put the whole of them together, two stand out in greater prominence than any of the rest, viz., Arkwright and Crompton.

Probably the former did more than any other Englishman to establish what is known as the Modern Factory System. He was not what one might call a brilliant man or great inventor, but he had the happy knack of appreciating and seizing upon what he knew was a good thing, and set about instantly to get all out of it that he could, and there are those who strongly affirm that he often got much more than he was entitled to.

However that may be, it can not be denied that he possessed eminent business qualifications, and these, coupled with other of his qualities, helped to make him exceedingly successful.

He first saw the light of day on December 23rd, 1732, in Preston, Lancashire, twenty-one years before his great rival and contemporary, Samuel Crompton. His parents could not possibly afford to give him any schooling, he being the youngest of thirteen. Apprenticed to the trade of barber, he became in time a first-rate man in that business. In 1760, when twenty-eight years of age, he left Preston and settled down in Bolton in Lancashire, setting up the business of barber and peruke-maker. The youthful Samuel Crompton would no doubt pay him many visits when in Churchgate, and little did he dream that the head he so often would undoubtedly use his skill upon was the one which would evolve by and by a machine which would amaze the then commercial world; but it was so. Another part of Arkwright's business, that of travelling up and down the country buying and selling human hair for wig-making, would put him *au fait* with almost every new invention and idea.

Richard's business card proves that he believed in advertising himself even as a barber.

Just about this time there was much excitement, especially in Lancashire, about the marvellous invention of Hargreaves, the particulars of which had now become known to the public. One of the first to appreciate the significance of this invention was Arkwright himself, so that it may reasonably be supposed that he would in good time know all there was to be known of the mechanism used by Hargreaves in his new method of spinning.

Later on, Arkwright became acquainted with a man named Highs of Leigh, another experimenter in spinning. The circle of his acquaintanceship also included Kay, a clockmaker of Warrington, who had assisted Highs on several occasions in his investigations.

At this time Arkwright's all-absorbing hobby was mechanics, and first one experiment and then another was made in rapid succession. Needless to say, his business of barbering suffered in the meanwhile.

From the first he turned his attention to an improvement of spinning cotton by drawing rollers. His efforts were crowned with success, and he ultimately blossomed into a knight, and was elected High Sheriff of Derbyshire. It is rather singular that he should be about the only one of the cotton-machinery inventors of this age who amassed a fortune; most of the others being but slightly removed from want in their last days.

There were many who claimed that they were the real and original inventors of this method of spinning by rollers, but there can be no doubt that to Arkwright alone belongs the credit for bringing these improvements to a higher state of perfection than they ever attained before.

At the present time, roller drawing is the great basis of the operations of modern spinning, wherever performed.

Not only is this the case in the final stages of production, but it is especially true of most of the preparatory processes, whether used for the production of coarse, medium or fine yarns.

As is well known, the great principle of drawing rollers is, that the cotton is passed through three or four pairs of rollers in quick succession, and attenuated by each pair in turn, each pair being made to revolve more quickly than the preceding pair. This identical process is repeated in machine after machine, until finally the bulk of cotton is reduced to a fine thread, of which, in some cases, it takes two or three hundred miles to weigh *one pound*. Even in what are termed medium numbers or counts of cotton yarn, there are from fifteen to twenty-five miles of thread in a pound avoirdupois, and more than *a thousand million pounds* of such yarns are spun annually.

The year 1767 found Arkwright entirely absorbed in his ideas of roller drawing, and he got the clockmaker Kay to journey with him to Nottingham, possibly thinking that what had been meted out to other inventors in Lancashire should not be repeated in his case. He here collected about him a number of friends, moneyed and otherwise, who helped in his evolution of spinning machinery.

A man named John Smalley of Preston found him the wherewithal to carry on his experiments first at Preston and later on at Nottingham. Certainly what he put up at Nottingham gave such promise of practical utility, that two experienced business men were led to join him in partnership, and the three of them, Need, Strutt, and Arkwright, very soon had mills built in Nottingham, Cromford and Matlock. The first-named mill was worked by horses, the two latter by water, hence the common name of *water frame*, given to the machines of Arkwright.

The gentlemen taken into partnership were able and qualified to give good sound advice and help to Arkwright, and about the middle of the year 1769 he took out a patent for his "*water frame*."

To use his own words, in his specification he "had, by great study and long application, invented a new piece of machinery, never before found out, practised or used, for the making of weft or yarn from cotton, flax, and wool; which would be of great utility to a great many manufacturers, as well as to His Majesty's subjects in general, by employing a great many poor people in working the said machinery, and by making the said weft or yarn much superior in quality to any heretofore manufactured or made."

No useful purpose could be served by reproducing Arkwright's description of the machine in question, but a picture of the actual machine is shown in Fig. 22.

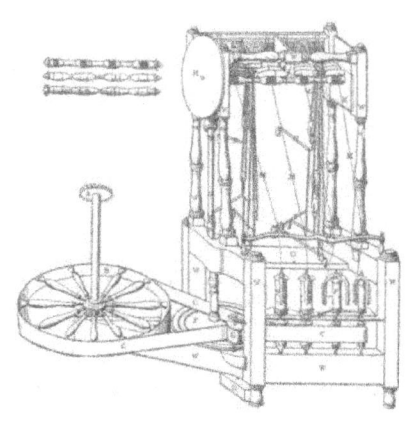

FIG. 22.—**Arkwright's machine (after Baines).**

The most important feature of the invention, of course, was the drawing out or attenuating of the cotton by rollers revolving at different speeds. But it was also essential that proper mechanism should be provided by which twist would be put into the yarn to make it sufficiently strong; and furthermore, it was necessary to arrange for the attenuated and twisted cotton to be automatically guided and coiled up or wound up into a convenient form. As we have seen, the drawing out of the cotton finer he accomplished by the Drawing Rollers originally invented by Lewis Paul, while for the latter purpose he successfully adapted the principle already existing in the Saxony wheel, used in the linen manufacture, with which he probably became acquainted during his residence at Preston.

It should not be forgotten that Hargreaves had introduced into the commercial world his Jenny, a few years anterior to Arkwright's water frame becoming so successful. These two machines were more or less in rivalry, but not perhaps to that extent which many would suppose. From the very first it was found that the frame of Arkwright's was much more suitable for warp or twist yarns, *i.e.*, the longitudinal threads of a cloth, whereas Hargreaves' machine was more adapted for the production of weft yarns, *i.e.*, the transverse threads of a cloth. Now it cannot be too strongly remarked that, at the present time, after the lapse of a century, the same state of things practically obtain in the improved machines of to-day; Hargreaves' machine being represented by the system of intermittent spinning upon the improved self-actor mule, while Arkwright's water frame is represented by the system of continuous spinning upon the modern Ring Spinning frame. While weft yarn is now almost entirely produced on the mule, warp yarns are in many cases now obtained from the Ring Frames, this latter system at the present time being greatly on the increase and daily becoming more popular.

The Carding Engine was greatly improved by Arkwright's many useful improvements, especially that of the Doffer comb, being entirely his own. The effect of this comb is fully described in the chapter dealing with manipulation of the cotton by the Carding Engine.

Paul was probably the first, in 1748, to invent the Carding Machine. His inventions seemed to hang fire until introduced into Lancashire, when they were adopted by a Mr. Peel, Arkwright and others. The chief defects, perhaps, of this machine was the absence of proper means for putting the cotton on the revolving cylinder and having it stripped when sufficiently carded. Hence the great value of Arkwright's stripping comb.

Some old Carding Engines which were used at this time are still in existence, though only used for museum purposes. As will have been gathered in a former chapter dealing with the manipulation of the cotton in the mill, between the Carding Engine and the final process of spinning there are other and important stages of preparation, and in these it is seen how in one respect Arkwright's method of drawing out cotton by revolving rollers was immeasurably superior to the travelling carriage of Hargreaves.

The strength of a rope is represented by its weakest parts, and the same may be said of yarn. There can be no doubt that one of Arkwright's greatest difficulties was to give an uniform yarn, and though he successfully launched his new machines he felt there was still much to be done in the direction of remedying yarn which was irregular in thickness and strength. In order to do this, he finally adapted his drawing rollers to what is now the modern drawing frame—a machine quite as largely used, and quite as necessary in present-day spinning, as it was a hundred years ago.

It was sought to make this machine do two things. (1) Several slivers of cotton from the Card were put up together at the back, and by means of four pairs of drawing rollers, were reduced to the thickness of one sliver (see the description in chapter vi.). It will be sufficient to say here that this method of doubling and drawing equalises the sliver of cotton by the combination of the thick places with the thin.

Doubling is now the reason of the uniformity of the yarns that are produced in such large quantities.

(2) The Carding Engine did not by any means lay the fibres of cotton sufficiently parallel to each other, and this process of parallelisation was fully accomplished by the front ends of the fibres being drawn forward more rapidly than their back ends by the drawing rollers revolving at different velocities. Mr. Baines says it was common to perform this operation until the finished sliver contained portions from *several thousand* carding slivers, but we think he would have been nearer the mark if he had said several hundred; although the higher number may be occasionally reached.

Yet again, in order to obtain a thread or yarn of sufficient fineness, it was found necessary to perform some of the attenuation of the cotton sliver as it left the drawing frame and before it reached the final spinning process. To this end, Arkwright adopted the Roving frame, in which the leading feature was again the celebrated drawing rollers. This machine made a soft and moderately twisted strand or roving, and if much twist had been put in, it would have refused to draw out finer at the spinning machine. Hence the means provided by Arkwright for the twisting and winding-on of the attenuated cotton on his spinning frame were utterly inadequate to cope with the soft loose roving, and as a matter of fact Arkwright never did see this problem satisfactorily solved.

He allowed, in his machine, the roving to fall into a rapidly revolving can which stood upright; the revolution imparting twist to the cotton. When this can was filled, it was carried to a winding frame, by which the roving was wound upon bobbins suitable for the spinning frame.

That Arkwright was unscrupulous in some of his dealings will soon be gathered if the various trials which he instituted to defend his so-called patents be carefully read, though it must be admitted that he possessed a most wonderful business capacity, and that he worked early and late, in pushing his ideas with the most tireless energy and determined perseverance. A glimpse of the nature of his early struggles is obtained when it is recorded that on one occasion his wife broke some of his first rude models, under the impression that he would starve his family by neglect of his legitimate business of barber. So incensed at her for this was he, that he ceased to live with her. Such were the defects of his early education and such his determination to learn, that at fifty he did not think he was too old to begin English grammar, writing and arithmetic.

That he succeeded in getting together a large fortune is now history. He died at the age of sixty on the 3rd August, 1792, at Cromford in Derbyshire.

Samuel Crompton.—Perhaps the greatest of the cotton-spinning machinery inventors was Samuel Crompton, who was born a few miles away from Bolton in a delightfully secluded and sylvan spot, "Firwood Fold," on the 3rd December, 1753. No story of the Cotton plant would be complete without mention of this individual, for wherever fine spinning machinery is practised there is a monument to the ingenuity, the skill and brilliant genius of Samuel Crompton. At a very early age he, along with his parents, removed into a much larger house still in existence and known as "The Hall ith Wood." This ancient mansion stands on a piece of high rocky ground and is distant from Bolton about 1½ miles. It was in this house that he invented his celebrated machine which he called "A Mule." At the present time one looks in vain for the Wood, but in the early days of Crompton's tenancy it was surrounded by a great number of very fine trees, hence the name "The Hall in the Wood" or "Hall ith Wood."

For some reason the Hall is being allowed to fall into decay, and at the present time is in great danger of collapsing. Several attempts have been made to buy the place and reclaim as much of it as possible and convert it into a museum, but as yet nothing has been done. It was built at two different periods: one portion of it, that of the "post and plaster work," being built probably in the 15th century, while the newer or later portion of stone was erected about 1648, for that date is inscribed on the porch.

The inside does not appear to have received much care or improvement. Originally the windows were much larger than at present. Pitt's window tax, long since repealed, was the direct cause for the reducing of the windows from their former proportions.

The illustration gives an excellent idea of its present-day appearance. The building is always an object of extreme interest to visitors to the locality, presenting even now a very picturesque appearance.

FIG. 23.—The Hall ith Wood, where the spinning mule was invented.

Very soon after the removal of the family to the Hall ith Wood, Samuel's father died. His mother, however, one of the best of women, filled the duties of head of the house with much success, and followed the laborious occupation of farming, and in her leisure moments, did what many housewives of her class did—carded, spun, and wove, in order to provide her family and herself with a decent livelihood.

She managed to give what might be termed under the circumstances a most excellent and practical education to her son Samuel; and it may be here remarked, that in many respects he was the exact opposite of his predecessor Arkwright. The latter was certainly a bustling, pushing man of business, while Crompton was a born inventor and recluse, and be it said also, as big a failure, as a business man, as could be well conceived. Of course Arkwright, as is well known, was the opposite of this.

The early youth of Crompton was identified with the great progress in the cotton industry of England, and, at fifteen or sixteen years of age, he was to be found assisting his mother during the daytime, while in the evenings he attended night-classes in Bolton, where he made great progress in mathematics. He was so good at the latter subject that he was called "a witch at figures."

It may be taken as perfectly natural that a man of the character, training and early associations of Crompton should turn to invention in connection with the cotton industry, especially since the beginning of his association with the trade there had always been a scarcity of weft for the loom which he and his mother operated.

The continual efforts of English weavers of that period to produce fine cotton goods to compete with those at that time largely imported from India, led to a great demand for fine yarns, and these the comparatively clumsy fingers of English spinners could not produce in a manner at all equal to the delicate filaments produced by the Hindoos.

Kay's invention of the fly shuttle, and the introduction by his son of the drop-box in the loom, had vastly increased the output of the loom, thus increasing the demand for weft and warp to feed it.

The inventions of Arkwright, Paul and others had certainly done much toward supplying this demand, but in Crompton's youth and early manhood the need of suitable weft was greater than ever. Mrs. Crompton was not long in hearing about the Jenny of Hargreaves, and determined she would get one for her son to work upon. This she did, and Crompton very soon became familiar with it and produced upon it sufficient weft for their own use. This he continued to do for seven or eight years, although he constantly had the truth forced upon him, that the yarn he was producing was neither as suitable for warps as that from Arkwright's water frame, nor at all adapted for the fine muslins then very much in requisition for ladies' dresses.

The manufacture of these muslins and of cotton quiltings was commenced in Bolton, Lancashire, by Joseph Shaw, when Crompton was about ten years of age; and from that time up to the present, no town in the world enjoys the same reputation for this class of goods as does Bolton.

With so contemplative and reflective a mind as Crompton's, and the many years of constant and, to a great extent, solitary occupation on Hargreaves' Jenny, it is not to be wondered at that Crompton's ingenious brain led him to devise some mechanism for improving the jenny on which he worked.

In 1774, therefore, he began those experiments which, after five years labour, resulted in the invention of the "New Wheel," or "Muslin Wheel," or "Hall ith Wood Wheel," as it was variously designated. The term "Mule" was of later application, owing to its comprising the essential features of both Arkwright's and Hargreaves' inventions. Because it was a cross or combination of the two it received the name of Mule, by which it is known to-day.

At the very time Crompton perfected his machine sufficiently to give it a practical test, the Blackburn spinners and weavers were going riotously about, smashing to pieces every jenny with more than twenty spindles, that could be found for miles around the locality, so that Crompton took elaborate pains to conceal the various parts of his new machine in the ceiling of his work-room at the Hall ith Wood in order to prevent their destruction.

Crompton's hopes and prospects were very bright at this time, as he had a watch costing five guineas expressly made for him, and just after the completion of his invention, he married one Mary Pimlott, at Bolton Parish Church, 16th February 1780. He was then but twenty-seven years of age, and his great invention, destined to revolutionise the cotton trade, was already an accomplished fact although practically a secret to the world at large.

When married, he and his wife set themselves assiduously to produce the finest strong yarn which his machine was so eminently adapted to spin. It did not take long for the good news to travel that fine yarn suitable for the production of muslins was being made at the Hall ith Wood. Hundreds of manufacturers visited Samuel to purchase, but many more came out of curiosity, if by any means they could see this wonderful machine. One individual is said to have hidden himself five days in the cockloft and, having bored a hole through the ceiling, feasted one eye at least by a sight of the marvellous mechanism which Crompton had invented.

Ballantyne records that as much as 14s. per pound was obtained for 40's yarn; 25s. for 60's, and for a small quantity of 80's, 42s. per lb.

At the time of writing the market prices for these are respectively, 7¾d., 9¾d., and 1s. 3d. per lb.

Crompton, however, was not permitted to enjoy his prosperity and monopoly very long, and here again may be noted the difference between him and Arkwright. While the latter extorted the full business profit from his inventions, the former suffered his ingenious machine to get out of his hands by promises not worth the paper on which they were written. His invention was not at all adequately protected by patent rights, and a number of manufacturers were allowed to use the Mule on their simple written promise to give him some remuneration. Long afterwards he wrote:

"At last I consented, in hope of a generous and liberal subscription. The consequence was, that from many subscribers, who would not pay the sums they had set opposite their names, when I applied to them for it, I got nothing but abusive language given to me to drive me from them, which was easily done, for I never till then could think it possible that any man (in such situation and circumstances) could pretend one thing and act the direct opposite. I then found it was possible, having had proof positive."

Another side of Crompton's character may be seen when it is stated he was an enthusiastic musician, and earned 1s. 6d. a night by playing the violin at the Bolton Theatre. Four or five years after the invention was known, he removed to the township of Sharples, where he occupied a farm-house called "The Oldhams," being probably induced to take this step in order to secure greater privacy.

A few words may very profitably be expended at this point in describing the main features of the machine shown in Fig. 24.

FIG. 24.—Crompton's spinning mule.

It has been remarked that Arkwright had already attained great success in the production of yarn by the extensive application of the principle of pulling out the cotton by drawing rollers. Hargreaves had also shown how to produce a thread by attenuating the cotton by means of a travelling carriage.

Crompton, however, laid the foundation of the present system of mule spinning by combining the essential features of the two machines and blending them into one.

He applied the principle of roller drawing in order to first attenuate the cotton, and he utilised the travelling carriage as a reserve power with which to improve the quality of the thread and draw it out finer.

It must not be supposed that his travelling carriage was identical with that of Hargreaves. On the contrary, it was a vast improvement upon it. Crompton put the twisting spindles into the travelling carriage and the roving bobbins he transferred to a fixed creel, and these conditions are invariably to be found in the self-actor spinning mule of to-day.

In Hargreaves' machine the rovings were placed on the travelling carriage, and the twisting spindles in the fixed frame behind, a position which has never been acceptable since that time for cotton-spinning mules. Here, however, a word may be said in favour of Hargreaves' disposition of the parts mentioned. The Jenny did not contain any heavy drawing rollers and roller beams, and it was probably best in his machine to have his crude roving creel to traverse and the twisting spindles to be in a fixed frame.

This disposition of the parts is even now to be found in most Twiner Mules, that is, mules used to double two or more single threads together without any process of drawing being applied to the cotton.

When Crompton applied the principle of drawing rollers, his ingenious mind saw that it would be best to let the rollers, roller beam, and roving creel be in a fixed framework on account of their combined weight and size, making it very difficult to move them about.

Crompton's great idea seems to have been to produce a better thread by his machine than could be given by other machines, and in this he admirably succeeded.

The mule being set in motion, the rollers first attenuated and then delivered the cotton to the spindle carriage. The latter, by the action of the hand and knee, was made to recede from the rollers just about as fast as the cotton was delivered to the spindles, or possibly at a rather quicker rate. Then, while the thread was still in a soft state, the rollers could be stopped and the threads pulled still finer by the continued recession of the spindle carriage from the rollers. Afterwards, when that length of thread was fully made, it wound on the spindles, and the carriage at the same time returned to the roller beam.

Thus each portion of thread was first subjected to the action of drawing rollers, as in Arkwright's machine, and then drawn still finer by the withdrawal of the travelling carriage, as in Hargreaves' Jenny.

Shortly after Crompton's invention was given to the public, it began to be improved in various ways. Henry Stones, a mechanic of Horwich, near Bolton, substituted metal drawing rollers for Crompton's crude wooden rollers, doubtless copying the idea from Arkwright's water frame.

All the mules employed at first were necessarily short; by that is meant they contained but few spindles, often 40 or 50 spindles. The biggest mule in Bolton in 1786 was said to contain 100 spindles. The preparation of the rovings for the mule about this time occupied the attention of Crompton, and he invented a Carding Engine which, however, did not attain very much success. Indeed it is said that one day so incensed was Crompton at the way he had been treated on account of his mule, that he took an axe and smashed his engine to pieces.

In 1791 Crompton established a small manufactory in King Street, off Deansgate, in Bolton.

In 1800 a subscription, promoted mainly by Manchester gentlemen, resulted in £500 being handed over to Crompton, one of the contributors for thirty guineas being the son of Sir R. Arkwright. With this money he was enabled to enlarge his business somewhat—one of his new mules containing upwards of 360 spindles and another 220 spindles. The mules were worked for many years, in fact, up to the sixties, when they passed into the hands of Messrs. Dobson & Barlow, the eminent cotton machinists of Bolton. One of the mules made by Crompton is shown in Fig. 24.

In the early part of 1812 an agitation for a government grant in recognition of Crompton's work made great progress. Mr. Perceval, the then Prime Minister, was proceeding to the House of Commons to move that a grant of £20,000 be made to Crompton, when he was shot by an assassin named Bellingham. There is no doubt, had this disastrous affair never happened and Perceval made his proposal, a grant much larger than was actually voted (£5000) would have been made.

There is no doubt that this grant was altogether inadequate, seeing that larger sums had been voted to other investigators and inventors about this time.

Owing to his lack of business ability, and to ill fortune combined, poor Crompton did not get out of this money what he might have done. Several ventures turned out altogether very differently than he expected. He became poorer and poorer, and was only protected from absolute want by subscriptions and assistance provided by his true friends in the trade, notably Mr. Kennedy, a Manchester manufacturer.

FIG. 25.—Portrait of Samuel Crompton. (*By the kind permission of W. Agnew & Son, Manchester.*).

At the age of 74 he died, 26th June, 1827. He was interred in Bolton Parish Churchyard, where a plain granite tomb sets forth the following:—"Samuel Crompton of Hall ith Wood, Inventor of the Mule, born 3rd December, 1753, died 26th June, 1827."

A noble monument of him is to be found standing on Nelson Square, Bolton, in front of the General Post Office.

CHAPTER IX.

THE MODERN SPINNING MULE.

The Self-Actor Mule.—In the preceding chapter there has been detailed the particulars of the invention of the "Mule" by Samuel Crompton. Since that event the mule has been the object of over a century of constant and uninterrupted improvement and development, especially in the details of greater or less importance.

The Self-Actor Mule of to-day represents and embodies the inventions of hundreds of the most intelligent men ever connected with any industry in the world's history. It is universally acknowledged to be one of the most wonderful and useful machines ever used. The actual operations of making a thread are however practically as left by Samuel Crompton over a hundred years ago. It is only in details of mechanism involved in making the various operations more perfectly automatic, and of greater size and productiveness, that the long line of inventors since Crompton's first mule was made, has been engaged.

To-day, such is the great size and wonderfully perfect automatic action of these machines, that they are found 120 feet long, while in width, over all, they may be 9 or 10 feet. Such a mule of this length would contain over 1300 spindles, each spinning and winding 64 inches of thread in about 15 seconds, and one man with two youths would be sufficient to give all the attention such a machine required.

Independently of a vast number of inventors of smaller importance, there are several names which stand out in greater prominence in the history of the developments of the mule. Among these names must certainly be placed, ahead of any others that might be named, that of Richard Roberts of Manchester, who succeeded in 1830, after about five years' application, in making the mule self-acting.

A good number of ingenious individuals had contributed more or less to this result between the dates of Crompton's and Roberts' inventions, and doubtless the results of the labours of these would be of great service to Roberts in his great task.

Indeed, several inventors had previously brought out what might be termed self-action mules, but it remained for Roberts to endow it with that constant and automatic motion which obtains to-day in practically the same form as left by him.

The special portion of mechanism with which his name is more especially identified, is what is denominated the "Quadrant." This is practically the fourth part of a large wheel, which is so arranged and connected that it performs almost exactly the same functions on a mule that Holdsworth's differential motion performs on the bobbin and fly frames.

To look at it, one would imagine it to be—what it really is—one of the simplest pieces of mechanism possible, yet the actions performed by it are complex and beautiful in the extreme. Later on, these actions of the Quadrant will be carefully examined.

FIG. 26.—Mule head showing quadrant.

The self-actor mule is an intermittent spinning machine, *i.e.*, it is not continuous in action, as are most machines used in the making of thread or yarn from the fibrous product of the Cotton plant. Take for instance the Carding Engine, and the bobbin and fly frames, as previously described. So long as these machines are working, practically all of the acting parts of the mechanism have a continuous forward motion.

This is by no means the case with the machine now under consideration, as many of the more important and principal parts move alternately in opposite directions, while other of the less important may revolve at one time, and be stationary at another.

What are called the medium counts of yarn contain say from 30 to 50 hanks in one pound avoirdupois; a cotton hank being equal to 840 yards, so that one pound of 40's yarn will contain no less than 40 × 840 yards or 33,600.

For such yarns as these, a modern self-actor mule would probably go through its cycle of movements four times per minute. For coarser or thicker yarns this speed might be increased, while for finer and better qualities of yarn the speed would be diminished.

Now as each succeeding "stretch" marks a complete cycle of movements and is a repetition of others, it will probably suffice if a brief non-technical description of one of these "stretches" or "draws," as they are termed in mill parlance, be given.

As in the bobbin and fly frames, the bobbins containing the rovings of cotton to be operated upon, are placed behind the mules on skewers fitted in a suitable framework of wood and iron called "creels," so as to allow the cotton to be easily pulled off and unwound without breaking. These rovings are guided to and drawn through three pairs of drawing rollers (see Fig. 27), which shows this very fully.

The chief difference between these rollers and those of the previously described machines being in the lessened diameters of the mule rollers, and consequently attenuating the cotton to a much greater extent. It is a truism well understood by those in the trade, that the finer the rovings are the better the raw cotton must be, and the more drawing-out they will stand in any one machine. One inch of roving put up behind the rollers of a mule spinning medium numbers would probably be drawn out into 9 inches.

FIG. 27.—Mules showing "stretch" of cotton yarn.

Nothing more need be said here about the action of the drawing rollers.

As the attenuated rovings leave the roller at the front, each one is conducted down to a spindle revolving at a high rate of speed; so quickly indeed, that there is no other body used in spinning which approaches it for speed.

It is quite a usual practice to have them making about 8000 revolutions per minute, and sometimes a speed of 10,000 is attained by them.

Assuming that a "Cop" of yarn (see Fig. 27), showing the cops on the spindles, has been partly made upon each spindle, the roving or thread from the rollers would extend down to the cop and be coiled round the spindle upwards up to the apex. The spindle would probably twist the thread for 40's counts twenty-three or twenty-four times for each inch that issued from the rollers, there being a well-recognised scale of "twists per inch" for various sorts and degrees of fineness of yarn.

Unlike the bobbin and fly frames, the roving or yarn is not wound on its cop or spindle as it is delivered, but a certain definite and regulated length of cotton is given out to each spindle, and fully twisted and attenuated before it is wound into a suitable shape for transit and for subsequent treatment.

To keep each thread in tension, therefore, as it is delivered from the rollers, the carriage containing the twisting spindles is made to recede quickly away from the rollers, a common distance for such movement being 64 inches. All the time the spindles are quickly revolving and putting twist into the rovings, thus imparting strength to them to a far greater degree than at any previous stage. Often the carriage is made to recede from the rollers a little quicker than the latter, the difference in the surface speeds between the two being technically known as "*gain*." The object of this carriage "gain" is to improve the "evenness" of the yarn by drawing out any thick soft places there may be in the length of thread between each spindle and the roller, a distance of 64 inches. It is a property of the twist that it will run much more readily into the thinner portions of thread than the thicker, thus leaving the latter capable of stretching out without breaking.

Arrived at the limit of 64 inches stretch (see Fig. 27), certain rods, levers, wheels and springs are so actuated that the parts which draw out the carriage and cause the rollers to revolve are disconnected, so that both are brought to a standstill for the moment.

In many cases the spindles at this stage are kept on revolving in order to put in any twist that may be lacking in any portion of the stretch.

Twisting being finished, the important operation of "backing off" commences.

It maybe at once explained that "backing off" means the reversing of the spindles; the uncoiling of a portion of the yarn from the spindles; and generally putting all the requisite apparatus into position ready for winding or coiling the attenuated and twisted rovings upon the spindles.

Here come now into action those most beautiful and ingenious applications of mechanical principles, the working out of which entailed so many years of arduous effort, and which rendered the mule practically self-acting and automatic.

By a most wonderful, intricate and clever combination of levers, wheels, pulleys and springs, aided by what is called a "friction clutch," the instant the spindles have ceased twisting the yarn, they are reversed in direction of revolution.

This reversal only occupies two or three seconds, and as the motion imparted to the spindles is very slow at this stage, the practical effect is, that a small portion of yarn is "*uncoiled*" from each spindle, sufficient to allow of two "guide wires" to assume proper and necessary positions for winding the attenuated threads upon the spindles.

These two wires are termed "faller wires," and while one is controlled by the cop-shaping mechanism and termed the "winding faller wire" the other simply keeps the threads in the requisite state of tension during "winding on" and is termed the "counter" or "tension faller wire." Both these wires can be seen in Fig. 28. During backing off, the "winding faller wire" has a descending motion, while the "counter faller" has an ascending motion, these being necessary for them to attain their proper positions for "winding on."

FIG. 28.—**Mule showing action of faller wires.**

The movement of these faller wires into proper position, and the uncoiling of a small portion of yarn from each spindle, are both brought about by the "backing off" motion, which formed an important part of Roberts' Mule. It may be remarked, however, that certain of the predecessors of Roberts had made great efforts in this direction, thus making the way much easier for his applications, which were entirely successful. When "backing off" is completed, all the necessary parts are in position for winding the 64 inches of thread just given out upon each spindle.

This practically involves three primary and most important operations. (1) The drawing-in of the carriage back to its original position. (2) The revolution of the spindles at a speed suitable for winding the threads upon the spindles as the carriage moves inwards. (3) The guiding of the threads upon the spindles in such a manner that a cop of yarn will eventually be formed upon each spindle, of such dimensions and shape as to be quite suitable for any subsequent processes or handling.

Taking these three important divisions in the order given, it may be said that the drawing-in of the carriage is effected through the medium of the "scroll" bands, which are attached to the carriage at one end, and to certain spiral scrolls or fusees at the other end. The scrolls being revolved, wind the cords or bands round them, so pulling in the carriage. There are usually two back scroll bands and one front band, the latter being a sort of check band upon the action of the other two. What is termed the "rim band" revolves the spindles during the outward traverse of the carriage.

The drawing-in of the carriage in a sense causes the other two operations to be performed. With respect to the second of these, viz., revolving the spindles and thus winding the threads upon them, it may be said this action causes what is termed the "Winding Chain" to pull off a small drum of six inches diameter, thus rotating the latter and thereby the spindles. Here, however, comes in now the action of the very beautiful and effective piece of mechanism, "Roberts' quadrant" (see Fig. 26). The winding chain just mentioned is attached to one extremity to the arm of the quadrant, and the peculiar manner in which the quadrant moves in relation to the winding drum gives the variable motion to the spindles that is required.

When commencing a new set of cops it may take about eighty revolutions of the spindles to wind on the 64 inches of thread to each spindle, representing one stretch. The bare spindle may be about a quarter of an inch in diameter, but it may finally attain a diameter of an inch and a quarter (*i.e.*, the cop upon the spindle). This cop will only require about twenty revolutions to wind on the 64 inches, which are only one-fourth of the revolutions necessary for the empty spindles. It is the action of the quadrant which gives this variation in speed to the spindles during winding-on.

But as has been pointed out previously, the quadrant imparts a "differential winding" motion to the spindles in two distinct and different ways, and the second motion is even more important than the first.

It is necessary for practical purposes that the cop of yarn should be built up of a conical shape in the upper part, as shown in the illustration. Now it must be obvious to the least technical of the readers of this story, that to wind a given portion of yarn upon the thin apex of a cone, will require a greater number of revolutions than would be necessary to wind the same length of yarn upon the base of the same cop. All the way between the apex and the base of the cone are also other varying diameters, and during each return movement of the mule carriage the thread is wound upon all the varying diameters of the cone in succession.

This implies the necessity for the revolutions of the spindles to a varying quantity all the time of the return or inward movement of the spindle carriage.

The quadrant gives this varying speed in a manner which is all but mathematically correct, any slight deviation from any such mathematical correctness being easily compensated for in other ways.

For the specific manner in which this quadrant works, the reader is referred to any of the recent text-books on cotton spinning.

The third primary and important operation, which takes place during each return movement of the carriage, is the guiding of the thread upon the spindles in a correct manner. This operation is closely associated, however, with the action of the quadrant.

That portion of a "self-actor mule" which guides the faller wires is termed the "shaper" or "copping motion." It consists of an inclined iron rail upon the upper smooth surface of which slides the "copping bowl," this being a portion of the mechanism which connects the rail with the faller wires. The rail rests upon suitable inclines termed "copping plates," whose duty it is to regulate the movement of the rail so as to allow for the ever-increasing dimensions of the cop during the building process. When the carriage again reaches its initial position, suitable mechanism causes all the parts to return in the position required for spinning.

Such is the complete cycle of movements of the "mule," each succeeding cycle being simply a repetition of the preceding. It will probably take such a mule as the one described about six hours to make a "set of cops," *i.e.*, one on each spindle, each cop being 1¼ inches in diameter and 7½ inches long.

Every fifteen seconds, while the mule is making a cycle of its movements, may be divided up approximately as follows: nine seconds for the drawing-out and twisting; two seconds for backing-off; four seconds for winding-on and resuming initial position.

A multitude of minor motions and details might be easily expanded into several chapters; in fact, more can be said about the mule than about any other spinning machine, but such detailed description would be out of place in this story.

All the motions just named are centred in what is termed the "Head Stock," this being placed midway in the length of the mule.

This head stock receives all the power to drive the various motions, from the shafting and gearing, and distributes it in a suitable manner to various parts of the machine.

It will have been observed by this time, that, as in the case of the bobbin and fly frames, the intricate and wonderful mechanism of the self-actor mule is not devoted to the formation of threads, but to the effective and economical placing of the threads of yarn, in the form of cops, after it has been spun.

FIG. 29.—Mule head showing "copping rail."

The spinning processes take place during the outward traverse of the mule carriage, the mechanism involved in this motion being comparatively simple. The really complicated and difficult motions being "backing-off," revolving the spindles "during winding-on," and the guiding of the spun threads upon the spindles during the winding-on process. It was the addition of these three motions by the later inventors which gave the mule the title of "Self-Acting."

CHAPTER X.

OTHER PROCESSES IN COTTON SPINNING.

The Ring Spinning Machine.—In a former chapter it was shown how within the space of two decades the three rival spinning machines of Hargreaves, Arkwright and Crompton were introduced, also it was pointed out, that Crompton's machines contained the best points of both of his predecessors. The mule did not immediately become the sole spinning machine. From the outset there was a close contest between the continuous spinning machine of Arkwright and the intermittent spinning machine of Crompton. It was not long, however, before the mule asserted its superiority over the water frame for fine muslin yarns, and for weft yarns. Eventually the water frame was relegated to the production of strong warp yarns, and later still it has come to be largely utilised as a doubling machine. As a matter of fact, it is contended by experts of the present day, that no machine ever made a rounder and more solid thread than the water frame, or flyer-throstle, as it has been called in its improved form.

FIG. 30.—Ring spinning frame.

During the last thirty years, a revolution practically in cotton spinning has been gradually brought about, and even to-day active developments are to be seen. The continuous system of spinning, which for a time had to take a second place, now appears to be again forging ahead, and looks as though it would supersede its more ponderous rival. Especially in countries outside England is this the case, for it is found that the method of ring spinning preponderates, and even in England the number of spindles devoted to continuous spinning is constantly increasing.

This change has chiefly been brought about by what may be termed a revolution in the winding and twisting mechanism of the continuous spinning machine itself.

Arkwright's flyer and spindle, after improvement by subsequent inventors, could not be revolved at anything like the speed of the spindle of the mule, and, in addition to this, the yarn had to be wound always upon the bobbin, very much after the style of the bobbin and fly frames previously described.

Experiments, however, were repeatedly made in the direction of dispensing with the flyer altogether, and some thirty years ago these unique spinning frames had attained very general adoption in the United States of America, where the comparative dearth of skilled mule spinners had furnished an impetus to improvement of the simple machine of Arkwright.

About this time, the attention of certain English makers being directed to the success of the new spinning frames in America, led to their introduction into England. But little time elapsed before they received a fair amount of adoption, but for many years they had a restricted use, viz., for doubling, that is, the twisting of two or more spun threads together, to form a stronger finished thread.

In this way, they were, strictly speaking, rivals of the throstle doubling frame more than the spinning mule.

By and by, however, the time came when the new frames began to be adopted as spinning machines, and to-day there are many English and foreign mills containing nothing else in spinning machines on the continuous system except these. In not a few mills in different countries, both types are found running.

A careful glance at the picture of this rival of the mule, will help in the following description of it:—

The flyer which is to be seen on the old Saxony wheel, and which was perpetuated in the celebrated machine of Arkwright, is entirely dispensed with, and all its functions efficiently performed by apparatus, simple in itself; it is yet capable of high speed and heavy production.

First of all, there is a vastly improved and cleverly constructed form of spindle, by which, in the latest and best makes, any speed can be attained which is likely to be required for spinning purposes.

Perhaps the apparatus which plays the most important part in performing the duties of the displaced flyer, is a tiny "traveller" revolving round a specially made steel ring about 2 inches in diameter.

The use of these two latter gives the distinctive names of "Ring-spinning" to the new system and "Ring Frame" to the machine itself.

In describing this system of spinning the creel of rovings to be operated upon, and the drawing rollers being practically identical with machines already described, little here is required to be said of them, but there is, however, a modification in the arrangement of the rollers which is referred to later on.

After leaving the rollers, a thread of yarn is conducted downwards and passed through the "travellers," which may be seen in the illustration, and then attached to the bobbin. The "traveller" is a tiny ring made of finely tempered steel. It is sprung upon the edge of the ring shown in the frame, and which is specially shaped to receive the tiny ring or traveller referred to.

The bobbin in this case is practically fast to the spindle—unlike any other case in cotton-spinning machinery—and it is therefore carried round by the spindle at the same rate of speed.

As the spindle and bobbin revolve, they pull the traveller round by the yarn which passes through it, being connected at one end to the bobbin and the rollers above forming another point of attachment. If the reader will look carefully at the illustration he will see how twist is put in the yarn. The joint action, then, of bobbin, traveller and fixed ring, is to put the necessary twist in the yarn which gives it its proper degree of strength. If no fresh roving from the rollers were issuing for the moment, the small portion of thread reaching from the rollers to the bobbins would simply be twisted without any "winding-on" taking place. As a matter of fact, the roving always is issuing from the rollers, and "winding-on" of the twisted roving is performed by the traveller lagging behind the bobbin in speed, to a degree equal to the delivery of roving by the rollers. It will be remembered that in the old flyer-throstle "winding-on" was performed by the bobbin lagging behind the spindle, a procedure which is impossible on the ring frame.

There is also an arrangement of the mechanism for guiding and shaping the yarn upon the bobbins in suitable form, the action being as nearly as possible an imitation of the mule.

For a number of years after the introduction of these frames, it was found that the threads often broke down owing to the twist not extending through the roving to the point where it issued from the rollers. This was eventually remedied by placing the drawing rollers in a different position, thus causing the thread running from the rollers to the traveller to approach more to the vertical; this constituting the modification which has just been referred to previously.

Another difficulty was experienced in the fact that during spinning the threads would sometimes fly outwards to such an extent that adjacent threads came in contact with each other, causing excessive breakage. This was technically termed "ballooning," and has been very satisfactorily restricted by the invention of special apparatus.

At the present time, therefore, a contest between the two rival systems of continuous spinning which were in bitter antagonism over a century ago, is waging a more fiercely contested fight than at any previous time.

As the case stands to-day, the mule is retained for nearly all the best and finest yarns as yet found; the most suitable for them, just as it was when Crompton got 25s. per pound for spinning fine muslin yarns on his first mule.

In many cases, also, yarn is specially required to be spun upon the bare spindle as on a mule, as for instance when used as weft and put into the shuttle of a loom. It is probably the very greatest defect of the ring frame that it can only, with great difficulty, be made to form a good cop of yarn on the bare spindle, although thousands of pounds have been spent on experimenting in that direction. How soon it may be accomplished with commercial success cannot be known, as a great number of individuals are constantly working in that direction. If it does come about, there can be no doubt that the ring frame will receive a still further impetus.

Even now, for medium counts of yarn it is much more productive than the mule, owing to its being a continuous spinner. Another vast advantage that it possesses is the extreme simplicity of its parts and work as compared with the mule. Because of this, women and girls are invariably employed on the ring frames, whereas it requires skilled and well-paid workmen for the mules.

The Combing Machine.—As compared with the Scutcher, the Carding Engine and Mule, the Comber is a much more modern machine. Combing may be defined as being the most highly perfected application of the carding principle.

The chief objects aimed at by the comber are:—To extract all fibres below a certain length; to make the fibres parallel; and to extract any fine impurities that may have escaped the scutching and carding processes.

It is worthy of note that although nearly all the great inventions relating to cotton-spinning have been brought out by Englishmen, the combing machine is a notable exception. It was invented a few years prior to 1851 by Joshua Heilman, who was born at Mulhouse, the principal seat of the Alsace cotton manufacture, in 1796.

Like Samuel Crompton—the inventor of the mule—Joshua Heilman appears to have possessed the inventive faculty in a high degree, and he received an excellent training in mathematics, mechanical drawing, practical mechanics, and other subjects calculated to assist him in his career as an inventor.

Heilman was the inventor of several useful improvements in connection with spinning and weaving machinery, but the invention of the comber was undoubtedly his greatest achievement.

He was brought up in comparatively easy circumstances, and married a wife possessing a considerable amount of money; but all that both of them possessed was swallowed up by Heilman's expenses in connection with his inventions, and he himself was only raised from poverty again by the success of the comber shortly before his death, his wife having died in the midst of their poverty many years previously.

After Heilman became possessed of the idea of inventing a combing machine, he laboured incessantly at the project for several years, first in his native country and subsequently in England. The firm of Sharpe & Roberts, formerly so famous in connection with the self-actor mule, made him a model, which, however, did not perform what Heilman required.

Afterwards he returned again to his native Alsace still possessed with the idea, and finally it is said that the successful inspiration came to him whilst watching his daughters comb out their long hair. The ultimate result was that he invented a machine which was shown at the great exhibition of London in 1851 and immediately attracted the attention of the textile manufacturers of Lancashire and Yorkshire.

Large sums of money were paid him by certain of the Lancashire cotton spinners for its exclusive use in the cotton trade. Certain of the woollen masters of Yorkshire did the same, for its exclusive application to their trade, and it was also adopted for other textiles, although Heilman himself only lived a short time after his great success.

It must be understood that the comber is only used by a comparatively small proportion of the cotton spinners of the world. For all ordinary purposes a sufficiently good quality of yarn can be made without the comber, and no other machine in cotton spinning adds half as much as the comber to the expense of producing cotton yarn from the raw material.

To show this point with greater force, it may be mentioned that the comber may make about 17 per cent. of waste, which is approximately as much as all the other machines in the mill put together would make.

Its use, however, is indispensable in the production of the finest yarns, since no other machine can extract short fibre like the comber. It is seldom used for counts of yarn below 60's and often as fine yarns as 100's or more are made without the comber. In England its use is chiefly centred in the localities of Bolton, Manchester, and Bollington, although there is a little combing in Preston, Ashton under Lyne, and other places.

Perhaps its greatest value consists in the fact that its use enables fine yarns to be made out of cotton otherwise much too poor in quality for the work; this being rendered possible chiefly by the special virtue possessed by the comber of extracting all fibres of cotton below a certain length. This of course has led to the increased production and consequently reduced price of the better qualities of yarn.

Reverting now to the Heilman Comber as it stands to-day, an excellent idea of the machine as a whole will be gathered from the photograph in Fig. 31.

There are usually six small laps being operated upon simultaneously in one comber. Each small lap being from $7\frac{1}{2}$ inches to $10\frac{1}{2}$ inches wide, being placed on fluted wooden rollers behind the machine, is slowly unwound by frictional contact therewith, and the sheet of cotton thus unwound is passed down a highly polished convex guide-plate to a pair of small fluted steel rollers.

Both the wooden and the steel rollers have an intermittent motion, as indeed have also all the chief parts of the machine concerned in the actual combing of the cotton. The rollers, during each intermittent movement, may project forward about $\frac{3}{8}$ of an inch length of thin cotton lap.

FIG. 31.—Combing machine.

By this forward movement the cotton fibres are passed between a pair of nippers which has been for the instant opened on purpose to allow of this action. Immediately the cotton has passed between the nippers, the feed rollers stop for an instant and the jaws of the nippers shut and hold the longer of the cotton fibres in a very firm manner.

The shorter fibres, however, are not held so firmly, and are now combed away from the main body of the fibres by fine needles being passed through them. The needles are fixed in a revolving cylinder and are graduated in fineness and in closeness of setting, so that while the first rows of needles may be about 20 to the inch, the last rows may contain as many as 80 to the inch, there being from 15 to 17 rows of needles in an ordinary comber.

The short fibres being combed out by the needles are stripped therefrom, and passed by suitable mechanism to the back of the machine to be afterwards used in the production of lower counts of yarn.

The needles of the revolving cylinder having passed through the fibres, the nippers open again and at the same time another row of comb teeth or needles, termed the top comb, descends into the fibres. The fibres now being liberated, certain detaching and attaching mechanism; as it is termed, is brought into action, and the long fibres are taken forward, being pulled through the top comb during this operation. Thus the front ends of the fibres are first combed and immediately afterwards the back ends of the same fibres are combed. During the actual operation of combing each small portion of cotton, the latter is quite separated from the portion previously combed, and it is part of the work of the detaching and attaching mechanism to lay the newly combed portion upon that previously combed. From a mechanical point of view, the detaching and attaching mechanism is more difficult to understand than any other portion of the comber, and it is no part of the purpose of this "story of the Cotton plant" to enter into a description of this intricate mechanism.

Sufficient be it to say that the combed cotton leaves the detaching rollers in a thin silky-looking fleece which is at once gathered up into a round sliver or strand and conducted down a long guide-plate towards the end of the machine. This guide-plate is clearly shown in the photograph of the comber, where also it will be seen that the slivers from the six laps which have been operated upon simultaneously are now laid side by side.

In this form the cotton passes through the "draw-box" at the end of the comber, and being here reduced practically to the dimensions of one sliver it passes through a narrow funnel and is placed in a can in convenient form for the next process.

When the combing is adopted, it precedes the drawing frame, which has previously been described, and the cans of sliver from the comber are taken directly to the draw-frame.

For intricacy and multiplicity of parts of mechanism, the comber is second only in cotton-spinning machinery to the self-acting mule, and is probably less understood, since its use is confined to a section of the trade. The latest development is the duplex comber, which makes the extraordinarily large number of one hundred and twenty nips per minute, as compared with about eighty-five nips per minute for the modern single nip comber. All this is the result of improvement in detail, as the principle of Heilman's Comber remains the same as he left it. It ought to be added that other types of comber have been adopted on the continent with some show of success.

FIG. 32.—Sliver lap machine.

Sliver Lap Machine.—Combing succeeds carding and is practically a continuation of the carding principle to a much finer degree than is possible on the card. The Carding Engine, however, makes slivers or strands of cotton, while the comber requires the cotton to be presented to it in the form of thin sheets. It therefore becomes requisite to employ apparatus for converting a number of the card slivers into a narrow lap for the comber.

The machine universally employed is termed "The Sliver Lap Machine," or, in some cases, "The Derby Doubler," and a modern machine is shown in the photograph forming Fig. 32.

In this case, eighteen cans are placed behind the machines, and the sliver from each can is conducted through an aperture in the back guide-plate designed to prevent entanglements of sliver from passing forward. Next each sliver passes over a spoon lever forming part of a motion for automatically stopping the machine when an end breaks. The eighteen slivers now pass side by side through three pairs of drawing rollers with a slight draft, and between calender rollers to a wooden "core" or roller. Upon this roller the slivers are wound in the form of a lap, being assimilated to one another by the action of the drawing and calender rollers.

Special Drawing Frame.—In order to have the fibres of cotton in the best possible condition for obtaining the maximum efficiency out of the combing action, it is the common practice to employ a special drawing frame between the card and the sliver lap machine.

As described elsewhere in this little story, the use of the drawing frame is to make the fibres of cotton more parallel to each other by the drawing action of the rollers, and to produce uniformity in the slivers of cotton by doubling about six of them together and reducing the six down to the dimensions of one. In the case under discussion the slivers from the card are taken to the special drawing frame and treated by it, and then passed along to the sliver lap machine as just described.

FIG. 33.—Ribbon lap machine.

Ribbon Machine.—Quite recently a machine has come slightly into use designed to supersede this special drawing frame. This new machine is termed the "Ribbon Lap Machine," and it may be described as a variation of the principle of the machine it is designed to supersede. The difference is this, that, whereas the drawing frame doubles and attenuates slivers of cotton, the Ribbon Machine operates upon small laps formed of ribbons or narrow sheets of cotton. By this treatment, the evening and parallelising benefits of the drawing frame are secured, with the addition of a third advantage, which may be briefly explained. The slivers, which in the sliver lap machine are laid side by side so as to form a lap, have a tendency to show an individuality so as to present a more or less thick and thin sheet to the action of the nippers of the comber. The latter, therefore, hold the cotton somewhat feebly at the thin places, thus allowing the needles of the revolving cylinder to comb out a portion of good cotton. When the Ribbon Lap Machine is employed, the slivers from the card are taken directly to the Sliver Lap Machine and the laps made by this machine are passed through the Ribbon Machine. Six laps being operated upon simultaneously by the rollers, are laid one upon another at the front so that thick and thin places amalgamate to produce a sheet of uniform thickness. The use of the Ribbon Machine is limited at present owing to its possessing certain disadvantages.

CHAPTER XI.

DESTINATION OF THE SPUN YARN.

Having initiated our readers into all the processes incidental to the production of the long fine threads of yarn from the ponderous and weighty bales of cotton as received at the mill, it remains for us to briefly indicate the more common uses to which the spun yarn is applied.

A very large quantity of yarn is consumed in the weaving mills for the production of grey cloth without further treatment in the spinning mill, except that the cops of yarn are packed in ships, boxes, or casks, in convenient form for transit purposes.

If for weft, the cops are forthwith taken to the loom, ready for the shuttle.

If for warp, then the yarn passes through a number of processes necessary for its conversion, from the mule cop or ring bobbin form, into the sheet form, consisting of many hundreds of threads, which are then wound on a beam.

Briefly enumerated, these processes are as follows:—

(*a*) The winding frame, in which the threads from the cops or spools are wound upon flanged wooden bobbins, suitable for the creel of the next machine.

(*b*) The beam warping frame, in which perhaps 400 threads are pulled from the bobbins made at the winding frame, and wound side by side upon a large wooden beam.

(*c*) The "slasher sizing frame," in which the threads from perhaps five of the beams made at the warping machine are unwound and laid upon one another, so as to form a much denser warp of perhaps 2000 threads, and wrapped on a beam in a suitable form for fitting in the loom as the warp or "woof" of the woven fabric. In addition to this, the sizing machine contains mechanism by which the threads are made to pass through a mixing of "size" or paste, which strengthens the threads.

In some cases this "size" is laid on the yarn very thickly, in order to make the cloth weigh heavier.

(*d*) After sizing comes the subsidiary process of "drawing in" or "twisting in," by which all the threads are passed in a suitable manner through "healds" and "reeds," so as to allow of their proper manipulation by the mechanism of the loom, to which they are immediately afterwards transferred.

In the production of cloths of a more or less "fancy" description, it is often required that the spun yarns shall be bleached and dyed before using, and to perform one or both of these operations efficiently, it is usual to reduce the yarn into proper condition by the processes of "reeling" and "bundling," although in comparatively few instances yarn is dyed in the cop form, while in a few other cases the raw cotton is dyed before being subjected to the processes of cotton spinning.

"Reeling" and "Bundling" are operations which are frequently necessary for other purposes besides those above alluded to, and may therefore be more fully described, as they often form part of the equipment of a spinning mill, and yarn is frequently sent away from the spinning mill in bundle form.

Reeling.—This is a simple but very extensively adopted process, in which yarn is wound from cops, bobbins or spools into hanks. It may be explained here that a cotton hank consists of 840 yards, and is made up of 7 leas of 120 yards each, while on a reel each lea is made up of 80 threads, a thread being 54 inches and equalling the circumference of the reel. Perhaps the most common size of reel contains at one time 40 spindles, and is capable therefore of winding 40 hanks of yarn simultaneously. The photograph in Fig. 34 shows a number of reels fitted for winding hanks from cops formed upon the mule.

The cops being put on the skewers, the end of yarn from each is attached to the reel or "swift" ready for starting. These reels may be arranged so as to be operated from shafting by mechanical power, or by the hand of the attendants.

FIG. 34.—Reeling machine.

Reeling is performed by women, and in our photo the attendant is seen in the actual operation of reeling.

A hank of yarn having been taken from each cop, the reel is stopped and closed up so as to allow of the ready withdrawal of the hanks.

Bundling Machine.—The Bundling press is solely intended to assist in the making up of the hanks of yarn into a form suitable for ready and convenient transit. In order to exercise a sufficient pressure upon the yarn to make a compact bundle, it is necessary for the framing to be of a very strong character, as will be especially noticed in Fig. 35.

FIG. 35.—Bundling machine.

The bundles of yarn made up on the bundling machine are usually 5 to 10 pounds weight, the latter being by far the more common size. The bundle shown in the yarn-box of our illustration is 10 pounds in weight and is practically ready for removal.

Before placing the yarn in the machine, several hanks are twisted together to form a knot, and these "knots" comprise the individual members of the bundle shown in the illustration.

In the sides of the yarn-box there are four divisions, through which are threaded as many strings, upon which may be placed cardboard backs. Then the knots of yarn are neatly placed upon the strings, and the cardboard and the strong top bars of the press securely fastened down. Certain cams and levers are then set in motion, by which the yarn table is slowly and powerfully raised so as to press the yarn with great force against the top bars. A sufficient pressure having been exerted, the bundle is tied up and withdrawn from the press, only requiring to be neatly wrapped in stout paper to be quite ready for transit purposes.

Sewing Thread.—A very large quantity of spun yarn is subsequently made into sewing thread. It is a fact well known to practical men that we have no means in cotton spinning by which a thread can be spun directly of sufficient strength to be used as sewing thread. For instance, suppose we wanted a 12's sewing thread, *i.e.*, a thread containing 12 hanks in one pound of yarn; it would be practically impossible to spin a thread sufficiently good to meet the requirements of the case. The method generally adopted is to spin a much finer yarn and to make the finished thread by doubling several of the fine spun yarns together in order to form the thicker final thread. For instance, to produce a 12's thread it is probable that 4 threads of single 48's would be doubled together, or say 4 threads of 50's, to allow for the slight contraction of the yarn brought about by twisting the single threads round one another.

In order to perform this doubling operation in an efficient manner for the production of thread, it is usual to employ two machines.

The first of these is shown in the illustration, and is termed the quick traverse winding machine. Here the cops from the mule, or the bobbins from the ring frame, are fitted in a suitable creel, as shown clearly at the front and lower part of our illustration. Each thread of yarn is conducted over a flannel-covered board which cleans the yarn and keeps it tight. Then each thread passes through the eye of a small detector wire which is held up by the thread and forms part of an automatic stop motion which stops the rotation of any particular bobbin or "cheese" when an end or thread belonging to that "cheese" fails or breaks, leaving the needles or detector wires. All the threads—from two to six in number—belonging to one "cheese" are combined to form one loose rope or thicker thread.

FIG. 36.—Quick traverse winding frame.

It ought to be explained that the term cheese is applied to the kind of bobbin of yarn which is formed upon this particular machine, one or two being placed as shown on the frame work.

Doubling Machine.—The machine just described does not put any twist into the thread, although twisting is a process which is absolutely indispensable for the proper combination of the several single threads so as to produce a strong doubled thread.

The twisting operation is therefore performed on the machine illustrated in Fig. 37, and termed the "Ring doubling machine."

In the creel of this machine are placed the cheeses formed on the winding machine, and the threads are conducted downward and usually under a glass rod in trough containing water, as the addition of water helps to solidify the single threads better into one doubled thread. From the water trough the threads are conducted between a pair of revolving brass rollers which draw the threads from the cheeses and pass them forward to the front of the machine. Here each doubled thread extends downwards and passes through a "traveller" upon the bobbin.

This machine is a modification of the ring spinning frame previously described and therefore does not call for detailed treatment at our hands.

The two machines are practically identical in principle, the chief difference being that in the doubler there are no drawing rollers, as the cotton is not attenuated in any degree at this stage.

Other differences consist in having larger "travellers" and "rings" and "spindles," and in a different kind of bobbin being formed.

FIG. 37.—Ring doubling machine.

At the doubling mill these threads are submitted to finishing processes, by which they may be polished and cleared and finally wound upon small bobbins or spools ready for the market, as seen in Fig. 2.

A fair proportion of the very best yarns are utilised in the manufacture of lace and to imitate silk. Such yarns are usually passed through what is termed a "gassing" machine. In this process each thread is passed rapidly several times through a gas flame usually emanating from a burner of the Bunsen type. The passage of the thread through the flame is too rapid to allow of the burning down of the threads, but is not too quickly to prevent the loose oozy fibres, present more or less on the surface of all cotton yarns, to be burned away. This process is somewhat expensive, as it burns away perhaps 6 pounds weight of yarn in every 100 pounds. This, however, is obtained back again by the increased price of the yarn. It is a property of the cotton fibre that it can be made to imitate more or less either woollen, linen or silk goods, and since cotton is the cheapest fibre of the lot it follows that a considerable amount of cotton yarn is used in combination with these other fibres, in order to produce cheaper fabrics. Embroidery, crocheting and knitting cottons, and the hosiery trade absorb a large amount of the spun cotton yarn; the latter being doubled in most cases in order to fit it for the special work it is designed to do.

In a modern spinning mill the ground floor usually contains the openers, scutchers, drawing frames, carding engines and bobbin and fly-frames. The upper floors are usually covered by mules and other spinning frames.

FIG. 38.—Engine house, showing driving to various storeys.

In the last illustration (Fig. 38) is shown one of the latest engines built for special work such as is required in a cotton mill. The huge drum, on which rest the ropes and which can be clearly seen in the picture, is divided into grooves. A certain number of these is set apart for the special rooms. The strength of the rope is known and its transmitting power is also known. When the power required to drive say the first storey or second storey is calculated, it becomes an easy matter to distribute the ropes on the drum as required. This engine is now at work in the Bee-Hive Spinning Mill, Bolton.